AF598106

Amazon Ecosystem - Past Discoveries and Future Prospects

Edited by Heimo Mikkola

Published in London, United Kingdom

Amazon Ecosystem - Past Discoveries and Future Prospects
http://dx.doi.org/10.5772/intechopen.111216
Edited by Heimo Mikkola

Contributors
Heimo Mikkola, Christopher Davis, Matea Stiperski Matoc, Zoran Stiperski, Tomica Hruška, Patricio Trujillo-Montalvo, José Emerson Souza, Cleinaldo de Almeida Costa, Gabriel Peixoto França, Nivaldo Alonso, André Luis Willerding

First published in London, United Kingdom, 2024 by IntechOpen
IntechOpen is the global imprint of INTECHOPEN LIMITED, registered in England and Wales, registration number: 11086078, 5 Princes Gate Court, London, SW7 2QJ, United Kingdom

British Library Cataloguing-in-Publication Data
A catalogue record for this book is available from the British Library

Additional hard and PDF copies can be obtained from orders@intechopen.com

Amazon Ecosystem - Past Discoveries and Future Prospects
Edited by Heimo Mikkola
p. cm.
Print ISBN 978-0-85466-269-2
Online ISBN 978-0-85466-268-5
eBook (PDF) ISBN 978-0-85466-270-8

For EU product safety concerns:
IN TECH d.o.o., Prolaz Marije Krucifikse Kozulić 3, 51000 Rijeka, Croatia,
info@intechopen.com or visit our website at intechopen.com.

Meet the editor

Heimo Mikkola obtained a Ph.D. from the University of Kuopio, Finland. He has more than 40 years of experience in sustainable development through working with growing responsibilities in international organizations, particularly the African Development Bank, European Union, United Nations, and the World Bank. He has visited 139 countries. In South America, he worked with the Food and Agriculture Organization of the United Nations in Colombia and Uruguay. He also visited Brazil, Chile, Costa Rica, Cuba, Curaçao, Guadelupe, Panama, and Venezuela in this capacity. Dr. Mikkola is an adjunct professor at Eastern Finland University and has been a visiting professor at three different Kazakhstan and Kyrgyzstan universities between 2010 and 2018. He has published more than 700 papers and books. He is also a book editor for numerous titles.

Contents

Preface

The Amazon region is the world's largest river basin and rainforest ecosystem. The vast area belongs to nine South American nations: Bolivia, Brazil, Colombia, Ecuador, French Guiana, Guyana, Peru, Suriname, and Venezuela. Amazonia has one of the largest amounts of surface freshwater and the highest average annual precipitation in the world [2]. Amazon River is the lifeline for the region, covering almost 40 percent of the South American continent, making it the largest river in the world in terms of water volume. Some new studies implicate that the Amazon River is also the world's longest river with a length of 6990 km. It is navigable by ocean steamers as far as Iquitos in Peru. For a long time, the vastness and inaccessibility of the river have protected the region, which is believed to have unparalleled biodiversity; one in five known animal species lives there. The Amazon jungle is home to one million indigenous people, some three million species of plants and animals, and billions of trees. The forest is a vital carbon store that slows down the advance of global warming. There have been many climate fluctuations during the last 55 million years, but never before have "the lungs of the world" been at greater risk than they are today due to uncontrolled fires, expanding agriculture and logging, and heavy industrial development in the forms of oil drilling, mining, and large hydroelectric dams. This book highlights a new risk, that of historic drought caused by El Niño. This book includes five chapters describing the anthropological, biological, and industrial problems facing the Amazon. It also maps the national, regional, and international political and human efforts to save the Amazon and presents new sustainable solutions to solve the current problems.

I want to express my best thanks to Publishing Process Manager Tonci Lucic at IntechOpen for his active and skillful cooperation during the preparation and publication of this book.

Heimo Mikkola
Eastern Finland University,
Finland

Section 1

Traditional Life

Chapter 1

Introductory Chapter: Severe El Niño Drought Affecting the Amazon Region

Heimo Mikkola

1. Introduction

Amazon basin has an area of 2.7 million square miles belonging to nine countries as follows:

Bolivia 7.7%, Brazil 58.4%, Colombia 7.1%, Ecuador 1%, French Guiana 1.4%, Guyana 3.2%, Peru 12.8%, Surinam 12.5% and Venezuela 6.1% [1].

The Amazon River is the widest and deepest river in the world and has by far the largest flow of water and drainage area. The Amazon and the Tocantins-Araguaia basins in the north account for 55% of Brazil's total drainage area. Some studies consider the Amazon to be the longest river in the world with a length of 6990 km and the Nile the second longest with 6850 km [2].

Three of the largest tributaries of the Amazon basin are born in Colombia: the Putumayo River (1700 km), the Negro or Guainía River (2000 km) and the Caquetá River (2200 km) [2].

The total incoming water is 2985.5 km^3/year, of which 29.5% (880 km^3/year) comes from Colombia (Japura, Negro and Putumayo), 1495.5 km^3/year from Peru (Amazon, Jurua and Purús), 550 km^3/year from Plurinational State of Bolivia (Madeira) and 60 km^3/year from Bolivarian Republic of Venezuela (Casiquiare) [2].

Casiquiare River forms a unique natural canal between the Orinoco and Amazon River systems. It is the waterway of the world linking together two major river systems. This phenomenon is called bifurcation. The area forms a water divide, more dramatically at the regional flood stage [3].

With all this water flow and the fact that the planet's biggest rainforest holds a fifth of the global freshwater, it is almost impossible to believe that drought, deforestation and unrelenting heat could suck it dry and nearly paralyze all water-related activities in the region [4].

2. El Niño

Normally the Amazon region has a dry season from July to December, and water levels in the rivers can drop several feet causing the beaches and sandbanks to appear. The year 2023 was, however, totally different due to El Niño [5]. The official term for this weather phenomenon is the El Niño-Southern Oscillation (ENSO) cycle which means the unusual warming of surface waters in the eastern tropical Pacific Ocean causing the Pacific jet stream to move south and spread further east [6]. El Niño was first noted by South American fishermen in the 1600s when the sea water was unusually warm in the Pacific Ocean. Because that phenomenon was strongest in December, the fishermen named it El Niño de Navidad ('Little Boy of Christmas'). La Niña ('Little girl' in Spanish) has the opposite effect – a cold event when trade winds are strongest and push warm weather towards Asia [7]. In the past, El Niño led to 1998 and 2016 global heat waves but in 2023 this climate pattern started earlier than usual reaching the peak power in winter. That exacerbated a historic drought in the Amazon region [5].

3. Drinking water

The region has in normal circumstances the most freshwater reserves in the world. So, there has been no need for a public water supply because people have taken their water directly from the river. Now there is a lack of water, which is a great paradox [5].

4. Emergency

A state of emergency was declared in 55 municipalities by the Brazilian authorities mainly because inability of boats to navigate the low-water rivers which has left hundreds of thousands of inhabitants completely stranded in the remote corners of Brazil [5].

5. Fisheries

Many families got their most wanted fish, namely pirarucú Arapaima gigas, from the next-door river before the drought but now this world's largest freshwater fish (often more than 100 kg) cannot swim in the river tributaries with less than 20 cm water. The remaining big fish are hiding in the lakes, but without a boat, fishing is impossible as is bringing the catch back home [5].

6. Forest fires

It has been noted that the deforesters have taken advantage of the situation by burning the forest to open up the land for new pastures. In September 2023 nearly 7000 forest fires were recorded making it the second-worst month on official data since 1998 [8]. And these deforesters are not indigenous people who are known to be effective in resisting forest loss. Rainforests managed by natives contain a high level of biodiversity and larger carbon density than state-managed forests [8]. It was the President Jair Bolsonaro who in 2019 started to weaken the land rights of the indigenous people. During his three years as President, more than 34,000 square km of rainforest vanished from the Brazilian Amazon, this being 52% more than in the previous three years [9].

7. Landslides

During the normal dry season, landslides are fairly common, but in 2023 the situation was much worse. The loss of rain has caused a sharp decrease in the flow of rivers. When the weight of water decreases, it stops supporting the land masses above it. Earth starts cracking causing collateral damage. People say that in Villa Arumã a landslide covered up half the town and 45 houses disappeared [5].

8. Political consequences

Brazil's President Luiz Inácio Lula da Silva included indigenous people within his administration in 2023 [8]. He also promised to bring illegal mining under control and to monitor forest loss better, as he did between 2004 and 2012 [9]. One can only hope that this unprecedented drought will not stop the glimmer of hope President Lula da Silva was giving for the Amazon. His predecessor Jair Bolsonaro opened the Amazon for land invasions, illegal mining, large-scale ranching and non-sustainable agriculture, all activities that were the major reasons for human-caused forest fires [8]. When Bolsonaro came into power in 2019 most law enforcement measures were stripped, science and environment agencies funding was cracked down and environmental experts were fired [9].

9. Power supply

Some large hydroelectric power plants had to be shot down due to a lack of water to run the turbines [5]. If the water situation does not improve soon, large areas of the region could start suffering from electric blackouts.

10. Subsistence agriculture

The lack of rain caused serious problems in subsistence agriculture which is the lifeline of most inhabitants [5].

11. Transport

The low water level of rivers is preventing navigation with any size of boats, which have been the only means to supply thousands of inhabitants in small municipalities in northern Brazil. People are running out of food and clean water pushing them to their survival limits. Authorities estimated that some half a million people have been affected by drinking water, food and medicine shortages in the area [5].

12. Water temperature

The highest water temperatures during the drought have been over 39 degrees Celsius in the Tefé Lake at the end of September 2023 [10]. This is far too high for many water-living creatures.

13. Wildlife

The most relentless drought in recorded history is impacting not only human beings but also putting wildlife at risk [8]. Brazilian Tefé Lake region, on the border with Peru and Bolivia, is particularly important for the Amazon River Dolphins Inia geoffrensis (**Figures 1** and **2**). There have been an estimated 1400 botos or pink river dolphins, as they are also named, but in September–October 2023 more than 125 dolphins were found dead representing between 5 and 10% of the population [10]. According to the scientists at the local Mamirauá

Figure 1.
Amazon River dolphins Inia geoffrensis. Photo: Silvio Battista Piotto junior.

Figure 2.
Boto or pink river dolphin begging for food in Brazil. Photo: Wikimedia common.

Institute, the extremely high-water temperature must be the explanation for deaths [5]. The institute is carrying out emergency operations to save the Amazon River Dolphin [11]. This water mammal is known for its echolocation, which allows them to navigate and find prey in the dark, muddy water [12] but even they suffer from the excess heat.

Further comments

The Brazilian authorities including the Air Force and the Navy have distributed clean water, non-perishable food, hygiene and health kits to most needy people. The non-governmental organizations have also organized aid in the form of water purifiers and gasoline for the little stoves to fry the stable diet cassava bread and cook fish [5].

Although the powerful drought shows no signs of abating, serious droughts will not last forever. Patterns of El Niño and La Niña continue usually not more than one year, but can sometimes last longer [6].

Amazonian people feel that in the future they should be more prepared for similar calamities by introducing solar panels for internet and mobile phone connections and developing family farming to give better autonomy to the communities [5].

This El Niño drought is an emergency which should get the maximum attention globally and should lead to effective action to prevent such warming and loss of nature (Mikko Pyhälä, Pers. Com.). The present government in Brazil is fully committed to crack down on deforestation rates in the country, especially in the Amazon region. Unfortunately, the El Niño drought will affect parts of the tropical ecosystem, threatening hundreds of thousands of people and animals depending on it.

Acknowledgements

It was Emeritus Ambassador Mikko Pyhälä who kindly alerted me on the Amazon region drought. Thanks for that. Wikimedia Common and Silvio Battista Piotto Junior are acknowledged for the photos.

Author details

Heimo Mikkola
Eastern Finland University, Finland

*Address all correspondence to: heimomikkola@aol.com

References

[1] Coca-Castro A, Raymondin L, Bellfield H, Hyman G. Land use status and trends in Amazonia. Available from: http://segamazonia.org/sites/default/files/press_releases/Land_use_status_and_trends_in_amazonia.pdf [Accessed: July 15, 2020]

[2] FAO (Food and Agriculture Organization of the United Nations). AQUASTAT Country Profile – Brazil. Rome; 2015. Available from: https://www.fao.org/aquastat/en/countries-and-basins/country-profiles/country/BRA [Accessed: June 16, 2023]

[3] Wikipedia. Casiquiare canal. Available from: https://en.wikipedia.or/wiki/Casiquaire_canal [Accessed: April 28, 2023]

[4] Ionova A, Andreoni M. Severe Drought Pushes the Amazon Rainforest to the Brink. The New York Times; 2023. Available from: https://www.nytimes.com/2023/10/17/climate/amazon-rainforest-drought-climate-change.html [Accessed: December 02, 2023]

[5] Royo GJ. We don't know what will become of us. Available from: https://english.elpais.com/climate/2023-10-7/severe-droughts-push-the-amazon-to-its-limit [Accessed: January 05, 2024]

[6] National Geographic. El Niño. Available from: https://education.nationalgeographic.org/resource/el-nino/ [Accessed: January 05, 2024]

[7] National Oceanic and Atmospheric Administration (gov.). What are El Niño and La Niña?. Available from: https://oceanservice.noaa.gov>facts [Accessed: January 05, 2024]

[8] Rainforest Foundation US. 2023 Amazon rainforest fires. Available from: https://rainforestfoundation.org/engage/brazil-amazon-fires [Accessed: January 06, 2024]

[9] Jones B. What Jair Bolsonaro did to the Amazon Rainforest? 2022. Available from: https://www.vox.com [Accessed: January 13, 2024]

[10] AP. More than 100 Dolphins Found Dead in Brazilian Amazon as Water Temperature Soar. El Pais; 2023. Available from: https://elpais.com/usa/2023-10-03/more-than-100-dolphins- found-dead-in-brazilian-amazon-as-water-temperatures-soar.html [Accessed: December 04, 2023]

[11] Galarraga GN. Chatting with Alligators in the Amazon Rainforest. El Pais; 2023. Available from: https://elpais.com/travel/2023-06-06/chatting-with-alligators-in-the-amazon-rainforest.html [Accessed: October 16, 2023]

[12] National Geographic. Amazon river Dolphin (Boto). Available from: https://www.nationalgeographic.com/animals/facts/amazon-river-dolphin [Accessed: January 06, 2024]

Chapter 2

Amazonian Animism: Natural World Annotations in Paleoindian Cave Art

Christopher Davis

Abstract

Cultures that first settled the Lower Amazon basin in Brazil entered rainforests and caves that were already inhabited, long ago, by owls, bats, frogs, and numerous other animals. The animal activities left natural patterns that informed early cultures about viable locations, navigable paths, and ample resources. Rock Art painted in the Monte Alegre hills of Pará, Brazil, on the banks of the lower Amazon River appears to have annotated some of these locations and natural patterns, not always as direct representational art, but sometimes as animistic and mnemonic symbolism. This chapter presents three ways in which the natural world was revered in animistic rock art at Monte Alegre. The first is through inspirations or decisions for where, in the landscape, to paint in red and yellow hues, which may have been triggered by areas where naturally-occurring red and yellow lichen circles grew on trees and rocks. The second is through "patron" animals drawn, and often personified, at the entrance of caves where the natural animal proliferates. The third is through ritual magic of animal drawings "touched" by painted handprints. Amazonian animism is based on a legacy of honoring nature's patterned relationships.

Keywords: rock art, phenology, South American prehistory, tortoise, volcanism

1. Introduction

An insufferably hot, noxious cave in the middle of the Amazon rainforest is not what paleoindian pioneers in the hills of Monte Alegre, along the lower Amazon River, expected to find over 13,000 years ago. What they *did* find was a cave distinct from all others nearby, a cave permeated with animism so potent that it is deadly to humans. Yet, in trying to fathom the cave's manifestation, paleoindian artists annotated it with a type of personhood, likening it to a burrow, and the cave's unusual warmth to that of the sun. This article focuses on the rock art depiction of a red and yellow tortoise (a cold-blooded animal that burrows to regulate its temperature) deeper inside the cave, which symbolizes a non-human social relationship of the setting sun to death and the underworld.

Earliest human occupation in the region dates to approximately 13,200 years ago [1]. Many of the surrounding caves in Monte Alegre are inhabitable today, and even during the terminal Pleistocene when people first arrived there, some of the caves were probably hospitable *prima-facie*. A few other caves no doubt needed to be cleared of their denizens first: snakes, spiders, bats, owls, capybaras, or even jaguars. But one cave in the region is uniquely inhospitable and bizarre to all others.

Caverna do Diabo is unusually hot, with pungent ammonia gas lingering in the air above a ground writhing with roaches. All outward appearances should have repulsed humans away from this cave. Both hospitable and animal-infested caves and rock shelters surround it in all directions; there is no shortage of "better" cave options within walking distance. And yet, paleoindian artists (though possibly singular) ventured daringly inside—long enough, at least, to draw several cryptic designs and symbols on its interior walls.

Some of the red ochre pictographs clearly depict awkward, disembodied facial expressions. Others resemble animals, and one is a geometric pattern. Some drawings are abstract or undecipherable. Most of the pictographs are near the entrance. However, one panel of rock art is much deeper in the cave, and it contains a few closely-spaced images that compose a scene.

The artists certainly planned their works, since the ochre was applied wet, not scrawled from any in-situ ochre obtainable nearby. The artists also risked their health, and possibly their lives, in order to draw the image so far inside the cave; why? Such an inhospitable cave holds no apparent practical or material gains. The cave entrance lacks human artifacts, offerings, or even potentially commemorative visitation handprints, which are so common to many of the other rock art sites in Monte Alegre. Nor does a person's entrance into the noxious cave induce them with altered states of consciousness (none that I nor my companions with me experienced), which might otherwise have been considered worth the risk of life in order to divine knowledge, power, or healing. No contextual artifacts, features, or clearings suggest this cave, nor its immediate outside vicinity, was frequented or revered. Additionally, the themed elements of the art neither imply practical use (no ritual hunting scenes) nor transcendent imagery (no psychedelic, hallucinogenic, entoptic, therianthropic, nor other supernatural effigies), only warning.

Theorizing why the tortoise "scene" is deeper in the cave, and theorizing why the artists went through the effort to draw it in the first place, is the focus of this chapter. Geological data indicate the cave held its properties long before humans entered the continent, so there is high confidence that the rock art reflects the artists' commentary on the cave's characteristics sans direct human participation or agency. In other words, this cave was regarded as "animated"—self-perpetuated with logical purpose.

Not only does the existence of the rock art demonstrate that Amazonian paleoindians were strongly motivated by a belief system not anchored to purely human interests and economic or subsistence practicalities, but also the pictographic composition is evidence for their strong urge to annotate the land in order to substantiate their epistemological understanding of the world. One that acknowledged cohabitation with disembodied but self-motivated forces.

2. Geology and geography

Caverna do Diabo is located on *Serra da Maxirá* hill (**Figure 1-left**), one of many elongated low-relief hills arranged in a ring that delineate the Monte Alegre geologic dome, in the Brazilian state of Pará. The dome formed millions of years ago from a long-dormant hotspot. Additionally, a tectonic northeast-southwest fault, called

Figure 1.
Serra da Maxirá hill (left photo). Burraco quente (right) showing the chemically leached black and white boulders on the surface of the hill above Caverna do Diabo. This area feels warmer than its surroundings, there is a faint malodorous smell, and no vegetation grows here due to the forces emanating from the cave beneath.

the Ereré Fault, intrudes the bedrock of the Monte Alegre hill ridge [2, 3] through a graben-like landmass shift over 2 million years ago. The *Ereré Fault* lies beneath *Serra da Maxirá*, alongside *Serra do Erere*, and it produces a sulfurous thermal spring at the ground surface several kilometers to the northeast.

Evidence of ancient wildfires in sediment layers from the Tertiary period, and earlier layers, were determined by several geologists to be provoked by tectonic factors, like volcanism, electricity discharges, spontaneous combustion, and friction between rocks [2]. The Tertiary period ended over 2.5 million years ago, so any residual geothermal activity would have been much more subtle to humans first entering the hills over 13,000 years ago. These residual conditions being self-perpetual, however, may have fueled animistic rationale of the landscape by early humans.

Maxirá hill is round at the base and it reaches an altitude of only about 150 meters [2]. There are not many tall trees that obscure visibility or sunlight, and vegetation is not overly dense. The cave entrance is relatively easy to reach, and it is the only known cave in this hill. The serene ambience around *Maxirá* hill, however, belies the lethality of its cave.

The cave's name, *Caverna do Diabo*, means 'cave of the devil' (or demon) in Portuguese. It is aptly named for its insufferable conditions produced from suspected geothermal activity that keeps the cave well over 100°F (~38°C) even at its entrance. The cave also continuously emits ammonia gas, probably due to the same geothermal activity interacting with biological decay. The conditions inside the cave have been continuously sustained for a long enough period of time that rocks on the ground above it have been chemically altered (**Figure 1-right**). Despite the dangerous conditions, however, the entryway was painted with red pictographs.

We can only speculate what the first people at Monte Alegre thought about this cave from the rock art they produced. The fact that there is rock art is itself an indication that they too found this cave to be noteworthy, because they apparently took significant risk to paint inside it. The smell of ammonia is evident from several meters outside the entrance, and the increasing temperature is felt at the entrance. Covering the mouth and nose with cloth only offers temporary relief because as sweat and moisture accumulate, the gas condenses into liquid ammonia and its stinging smell burns the eyes, nose, and throat. The longer anyone remains inside, the greater the risk of liquid ammonia condensing into the lungs, and presumably the same would be true of the eyes. However, roaches and hornets endure the boundary zone at the entrance, and bats might be nesting inside.

3. Rock art and cave contexts

The rock art here is not plentiful, but they retain a wet or fresh appearance. The most visible and prominent pictographs were drawn on a jutting outcrop from the wall (**Figure 2**), which is about 1 meter high (~4 feet tall), just inside the entrance. All the drawings are visibly red, but Dstretch photo enhancement reveals yellow pigment too, mostly in the background, as though the surface was primed in yellow paint followed up by drawing the red pictographic image. Priming the background is

Figure 2.
Caverna do Diabo rock art. Unaltered photo of the entryway outcrop with the most prominent pictographs (top photo). The diagonal left side of the outcrop with geometric designs (bottom left photo enhanced with Dstretch lds to highlight red and yellow colors). The lower portion of the front outcrop showing two heads possibly of an owl and a bat (bottom center photo-Dstretch lds). The upper portion of the main outcrop depicting the 4-limbed creature with three torso dots (bottom right photo-Dstretch lds).

a suspected practice at other rock art sites in the hill ridge too [4]. This artistic trend again enforces the inference that the rock art was prepared and executed with premeditated intent.

The largest and most prominently placed drawing at the entryway of *Caverna do Diabo* is what appears to be a four-legged creature with a tail and a half oval torso that contains three dots in a curved pattern. The head of the creature is unidentifiable, but connected to one of its forelimbs is a circle with a dot in the center. Another encircled dot is also drawn near, but separate from, the creatures' hindlimbs. It is uncertain whether this creature is terrestrial or avian, insect or animal, nor can any of those possibilities be entirely ruled out. The three torso markings, however, were probably considered an identifiable detail by the artists and their community in the past.

Beneath the four-limbed creature are two disembodied non-human heads. The higher head is circular, with small circles forming the eyes, and a Y-shape above and between the eyes, which terminates on top of a single horizontal mouth line. Compared to other pictographic human and bird faces in these hills, this fits closer to an avian face—possibly an owl or hawk. The lowest pictographic head has a cleft forehead within a horizontally ovoid shape containing two small circles for eyes, and a narrow horizontally ovoid mouth. This head appears to represent a bat, although identification of both heads are not certain or indisputable.

Painted onto the surface of a diagonal outcrop just left of the previous one are some complex geometric lines in red pigment, and yellow pigment again appears to be in the background. What is discernible is an "X" shape to the left of a relatively square shape that encloses a diagonal slash (\). Other crisscrossed curvilinear lines appear above and to the left of the discernible ones. Square pictographs enclosing slashes and X shapes do also appear at two sites in Monte Alegre as well, most notably at *Painel do Pilão* in a large grid-like pattern of potential sky-related tally marks suggestive of a calendar.

The lower portion of the diagonal outcrop appears to have been broken off, although no fragments were visible on the ground underneath it. The broken portion of the outcrop also appears to have resulted in the removal of parts of the pictographs, which at least suggests the paintings predate detachment of the lower wall.

Just a bit further inside the cave, on the main back wall of the corridor entryway, is a red pictograph image that is only rendered discernible with Dstretch photo enhancements (**Figure 3**). Depicted near the top of the photo is an expressive face (without an enclosing head design). One eye is encircled while the other appears to be a mere dot. Above the encircled eye is a long arch that appears to be an eyebrow; the dot eye has

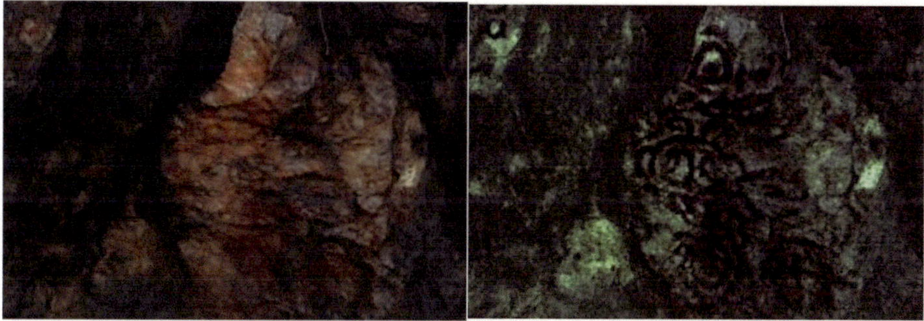

Figure 3.
Lower wall pictographs inside Caverna do Diabo. Unmodified photo of the pictographs (left) and Dstretch yre color enhanced photo to highlight yellow red and white (for increased contrast) colors of the rock art.

a short, but also probably faded, eyebrow. There is a short vertical line between the eyes, representing a nose. Beneath the eyes and nose are longer horizontal mouth lines that are hard to ascertain, but appear to bear a snear or grimace. Overall, the image appears to be a human facial expression.

Beneath the expressive face is a peculiar "baby" face framed in a horizontally ovoid "head" that has two ovals attached to either side, as though they are ears. The head contains encircled eyes that appear to be looking up and to the left. What makes the face peculiar are two lines sprouting from the center of the head, which form outward-turned hooks. These appear more like antennae than hair. Sprouting from beneath the head is a segmented curved J-shaped polygon that bears a resemblance to a worm or snake (or maggot?) torso. There might be more to the torso, since there are more red lines and marks, but they are too unfamiliar to be discernible.

The "baby" face is potentially looking at another unenclosed face above and to the left, given indication by a single circle "eye" with an upward arched "eyebrow" and a horizontal line "mouth." Although this face is at the edge of the photo, the rest of the face does not exist. The partial face was drawn on a projected portion of the wall, with the contours of the wall forming a lower jaw line. Not much else can be said about the significance of these drawings because the facial expressions and worm-like shape are unique in comparison to the rock art elsewhere in Monte Alegre.

Maxirá hill is notably the westernmost hill that contains rock art on the Monte Alegre hill ridge. Ethnography from other circum-Caribbean cultures indicate that some tribes refer to the land in the west, or particularly behind mountains, as the land of the dead [5–7]. Based on this context, the paintings in *Caverna do Diabo* might have death-themed, or ancestral-spirit significance, especially the "baby face" image, which could symbolize a human soul, or worms/maggots that feast from a corps, or perhaps both simultaneously.

However, the uncanny nature of the pictographs might also possibly reflect a supernatural understanding of the cave. The drawings bear some human resemblance in order to communicate a human warning or reaction. The "forces" at play in this cave are therefore bestowed animated *human-like* qualities [8] based on knowledge of how they affect humans, not based on how they affect (or appear not to affect) animals in the cave.

4. The Tortoise in the Devil's Lair

The next image from *Caverna do Diabo*, which is the focus of this chapter, was only discovered after image enhancements were done on a "blind" photo, having no knowledge that anything was inside the cave (**Figure 4**). To avoid prolonged exposure to the ammonia gas, the photo was taken blindly from just inside the corridor entryway, with the flash on, and using high film sensitivity (ISO 3200). The photo reveals a large boulder on the ground deeper inside, approximately 7–10 meters in the gallery of the cave. Drawn in red and yellow on the side of the boulder that faces the entryway is the appearance of a turtle/tortoise standing to the left of a plant (bush? tree?), and both are on top of a horizontal line that slants upward at either end (a sled?).

Although the boulder looks as though it could have possibly fallen from the wall or ceiling, such an assessment cannot accurately be made from a single photo taken from a distance. The pictograph appears to have been drawn within the space provided by the contours of the boulder, not drawn and then broken off as with the diagonal outcrop containing the geometric drawings.

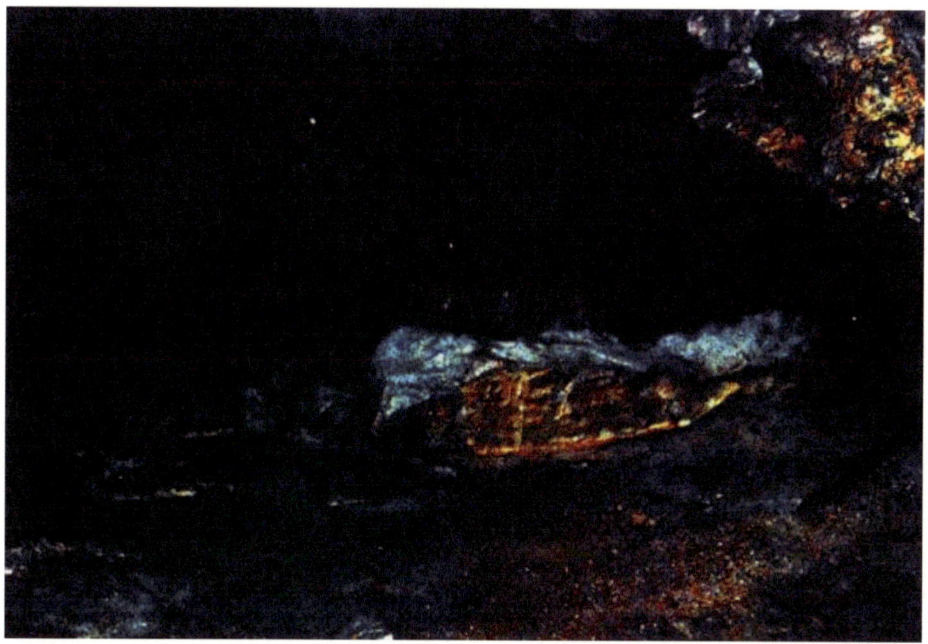

Figure 4.
Tortoise plant and sled pictograph. Photo taken "blind" inside Caverna do Diabo, revealing a red and yellow pictograph painted on a giant slab of rock on the ground. The photo was Dstretch lab enhanced, which enhances contrast and all colors.

This turtle/tortoise pictograph is unexpected, especially being found inside a cave. Turtles are cited, albeit rarely, as entering caves in the *Paituna* hill about 4 km south of *Maxirá*, which is closer to the Amazon river [2]. Tortoises are less known to inhabit caves, and the rare instances when they were found to do so [9], it was theorized to be a strategy for cooling their bodies down from the hot sun, which contradicts conditions found in this cave.

Turtles/tortoises are also not frequently depicted in the rock art, although one notable example may have been found at the grid pictograph at *Painel do Pilão* [4]. Despite the rarity of pictographic representation, turtles and tortoises both were very commonly sought and eaten, as has been indicated in archeological contexts spanning from paleoindian to historic-period excavation layers [10–12]. Culturally, they are very commonly featured in myths, particularly myths about the sun as represented by the Amazonian *jabuti* tortoise.

So why would there be a turtle/tortoise rock art image in a cave? To seek answers, it would first help to know which type of *chelonia* it is. Unfortunately, the photo, and perhaps the drawing, is not that detailed. However, the image enhancements do indicate that the shell is somewhat high arching like that of a tortoise, segmented with many small sections, and that the animal image is composed of both red and yellow pigment. The aforementioned *jabuti* now seems more plausible, though not exclusive.

The plant would be another important clue, but again the image is not detailed enough to determine the type of plant depicted. I had considered that the plant indicates a terrestrial habitat, but the Amazon has equally important aquatic and semi-aquatic plants (like *buriti* palm trees).

The "sled" could be another important clue because the animal is not "animated" by itself. Instead, like the plant, the tortoise/turtle appears rooted. Motion is implied instead by the sled. If this is indeed the case, the tortoise stands in as a symbol for something else that moves—an inanimate object imagined to be animate. The *jabuti* now seems arguably the best candidate, and it possibly symbolizes the sun. The next question to address is why the sun would be depicted in a cave?

Charles Hartt [13] further described other *jabuti*/sun analogy myths told among Tupí communities, like how the tortoise escaped the jaguar by burrowing into one hole and escaping out another. In this tale, Hartt applied the metaphor to the sun descending in the west to rise in the east the next day [13]. This metaphor implies that a red and yellow turtle pictograph, if representative of the sun, might be drawn in a cave as a metaphor for the underworld where the sun entered at night. This cave itself, being warm and powerful, was perhaps imagined as an underworld entrance for the sun to descend below the horizon in the west (relative to the other hills on the ridge). Therefore, the tortoise pictograph, as understood through animism, does not just explain, but also informs the viewer of the phenology of the cave—its power manifesting from the underworld (geothermal or tectonic forces from deep below the surface) as potent as the sun.

Phenology is the study of repeated natural patterns in relation to climate, plants, and animals. Ancient people engaged in a type of phenology as pioneers of new landscapes. Devoid of any human presence, they would have witnessed lands etched with animal paths and geological processes. South American megafauna still alive during the terminal Pleistocene when people first trickled into the Amazon were capable of drastically altering the landscape simply by moving, burrowing, climbing, and accumulating favored plants by their digestive habits. Phenologies offer a strong impetus for animistic beliefs that were meant to understand, or at least make note of landscape patterns pre-existent to human arrival.

5. Discussion

Animism is often assumed to be the foundational belief system of all religions [14]. The term *animism* means to endow with life-like movement emanating from disembodied spirits, neither benevolent nor malevolent (or both), that inhabit the world in locations like forests, lakes, springs, caves, and so on [15]. However, it is more than a set of beliefs. Animism is an entire perspective that anthropologists ascribe to nearly all hunter-gatherers before the Neolithic age about 10,000 years ago; and it still persists among many cultures today.

Animism as a perspective of relating the environment to the way aspects of it might be connected in a social sense has been termed 'relational epistemology' by Nurit Bird-David [16]. He sees animism as an epistemology that makes sense of an unpredictable or threatening world by relating everything in an interlocking relationship. The epistemology is rooted in predictions and explanations for how inanimate objects move, based on knowledge of how animate objects move [8].

Because we as humans move in patterns and make decisions based on motives, we imagine all patterns and movements are done in similar fashion, satisfying a will or purpose. Myths are often told from this paradigm. The problem is that we often cannot imagine a will or purpose beyond a human one. So, although the *tortoise-as-sun* myth borrows the reptile's behavior to explain the sun's behavior, the pictographic scene rests atop a (human) sled to rationalize the sun's mysterious motion.

The animism possibly associated with *Caverna do Diabo* is the idea that dangerous places contain powerful forces. The cave contains rock art, whereas numerous other more hospitable caves contain none. This cave could not have been inhabited by people, but it is at least partly inhabited by animals. The creature images on the right side of the entryway outcrop possibly identify animal denizens of the cave (perhaps wasps and roaches in addition to bats and owls?). However, the geometric shapes indicate human notation.

If the outcrop with geometric designs was indeed intentionally detached by humans, the geometric designs further support Bird-David's [16] aforementioned 'relational epistemology.' The cave certainly fits an "unpredictable or threatening environment," and so the animated landscape here would be rooted in the explanation of what makes this cave (and not others) so dangerous.

Would the ancient artists have surmised that volcanic activity over 2 million years ago left residual geothermal activity? Probably not, but animism renders such ponderings nearly trivial. Instead, the area was perceived as powerful, and therefore should be dealt with cautiously, except in times when powerful items are needed. If detachments were indeed taken from the outcrop, they quite possibly would have been perceived as being more powerful.

Intentional wall removal was discovered at a nearby rock art site, *Painel do Pilão*. However, the majority of rocks recovered from excavation directly underneath the wall at that site did not have pictographs on its surface. Pursuit of silicified sandstone suitable for knapping is one proposed theory for the *Painel do Pilão* site, the other proposed theory—particularly unique to the evidence unearthed at *Painel do Pilão*—was intentional modification of the landscape [4]. In either theory, animism is the implied logic behind wall removals.

Although never having been excavated, *Caverna do Diabo* lacks most of the associated material and pictographic context found at *Painel do Pilão*. Perhaps detached wall fragments from the geometric design at *Caverna do Diabo* were also taken away from the site by humans. If this is true, one could argue that despite the danger of the cave, venturing inside served a practical purpose. However, even if true, the practicality would have been minimal, and animistic beliefs remain the highest probable motivation to do so. Myths retained among other Amazonian cultures support this conclusion, and they often credit one or another phenology as providing culture-sustaining knowledge.

There is a myth, told among the Craho culture, of a hunter who went into the hole of an armadillo and found a bunch of peccaries living there, and therefore this is how hunters know where to find peccaries today [17]. Similarly, there are several versions of another Gê myth told by the Craho, Cayapo, Ramkokamekra, and Apinaye, where a hunter was pursuing an armadillo that fell into a hole that led to a large plain below. Young members of the hunter's tribe got rope and climbed down the hole (in some versions they climbed down a Buriti palm tree) and this is how people descended from the sky to populate the earth [17].

In both of these myths, the animistic conclusion drawn is that the events of non-human actors are decisive for human destiny. But perhaps this conclusion is somewhat cynical and hasty. Only if myths were simply meant to explain human serendipity (or suffering), then the cynical conclusion would perhaps be correct. However, if myth is meant to convey non-human patterns, and even a bit of history about non-human relationships too, then myths remain perpetually informative about the natural world, which helps us put that knowledge to good use.

In the terminal Pleistocene period, paleoindians pioneering into South America as the first humans might have discovered trails formed from *Neochoerus*, the 6-foot-long megafauna capybara, or caves widened and smoothed by *Megatherium*, the ground sloth that could grow up to 3.5 meters long [18]. Even the armadillo-like megafauna, *Glyptotherium*—with a weight and size comparable to a small car—may have burrowed underground holes readily useable by early human pioneers [18]. Paths which when followed, may have brought paleoindian hunters upon unsuspecting prey. That is, after all, implied by the Craho myth of the armadillo mentioned above.

The Amazon rainforest is teeming with life that no doubt spawned countless phenologies. Here, not just megafauna, but also owls, bats, frogs, and even lichen, lianas, and mangroves formed natural patterns that marked the land long before human arrival. These marks helped inform early cultures about viable locations, navigable paths, and ample resources. Some of these phenologies were perhaps translated through animism presented in rock art, others were presented through myth or both.

6. Summary

The case laid out by the tortoise pictograph in *Caverna do Diabo* is that Amazonian paleoindians entered a landscape that had already been set in motion by forces and animals, wherein people had to conform to it. Phenologies and non-human personhoods often aided the success of ancestral human pioneers in new lands. Primordial patterns, like that of a noxious hot cave, continuously manifest themselves with or without the presence of humans. Animistic cultures sometimes attempt to explain, but always find ways to annotate, them in myths and rock art that recount non-human relationships through the metaphoric relationships of animals and objects as though they were people with social relationships. Amazonian animism particularly emphasizes a legacy of honoring nature's patterned relationships. It was constructed from observed natural associations set within recognizable social relationships. The reasons are possibly for ease of memory, and perhaps to aid in recognizing and reinforcing interactions that accompany desired outcomes. The habits of animals and characters in myths may seem imaginative at first glance, unless we consider that the information, they preserve animates relationships within the natural world, sometimes wholly absent of human agents, except perhaps as third-party observers.

Author details

Christopher Davis
McHenry County College, Chicago, IL, USA

*Address all correspondence to: christopher.davis@fulbrightmail.org

References

[1] Roosevelt AC et al. Paleoindian cave dwellers in the Amazon: The peopling of the Americas. Science. 1996;**272**(5260):373-384

[2] Milhomens A, Daniela SN, Liliana S, Wagneide R, Fernando F, Amado M, et al. Plano de manejo do parque estadual monte alegre (PEMA). In: PROECOTUR (Programa de Desenvolvimento do Ecoturismo na Amazonia Legal) editors. Belem: Ministerio do Meio Ambiente; 2008

[3] Vizeu R, Almeida C. Relatório de Aspectos Abióticos da Região do Parque Estadual Monte Alegre. Brazil: Mimeo; 2006

[4] Davis CS et al. Paleoindian solar and stellar pictographic trail in the Monte Alegre hills of Brazil: Implications for pioneering new landscapes. Journal of Anthropology and Archaeology. 2017;**5**(2):1-17

[5] Guchte MVD, Brewer D, editors. Masquerades and Demons: Tukuna Bark-Cloth Printing. Urbana, Illinois: Board of Trustees of the University of Illinois; 1992

[6] Léry JD. History of a Voyage to the Land of Brazil, Otherwise Called America. Berkeley, CA: University of California Press; 1992

[7] Southey R. History of Brazil: Part the First. London: Longman, Hurst, Rees, and Deme, Paternoster row; 1810

[8] Wilson RA, Keil FC. The MIT Encyclopedia of the Cognitive Sciences. Cambridge, Massachusettes: The MIT Press; 1999

[9] Bittel J. Cave-Dwelling Giant Tortoises are a Big Surprise, in National Geographic. Washington D.C: National Geographic Society; 2016

[10] Roosevelt AC. Amazonia and the anthropocene: 13,000 years of human influence in a rainforest. The Anthropocene. 2013;**4**:69-87

[11] Roosevelt AC. The lower Amazon: A dynamic human habitat. In: Lentz DL, editor. Imperfect Balance: Landscape Transformations in the Precolumbian Americas. New York: Columbia University Press; 2000. pp. 455-492

[12] Roosevelt AC. Resource management in Amazonia before the conquest: Beyond ethnographic projection. In: Balee W, Posey DA, editors. Resource Management in Amazonia: Indigenous and Folk Strategies. New York: New York Botanical Garden; 1989. pp. 30-62

[13] Hartt CF. Amazonian Tortoise Myths. 1875. Available from: http://catalog.hathitrust.org/api/volumes/oclc/67400385.html

[14] Eliade M. Shamanism: Archaic Techniques of Ecstasy. Princeton, New Jersey: Princeton University Press; 1964

[15] Helvenston PA, Hodgson D. The neuropsychology of 'Animism': Implications for understanding rock art. Rock Art Research. 2010;**27**(1):61-94

[16] Bird-David N. "Animism" revisited: Personhood, environment, and relational epistemology 1. Current Anthropology. 1999;**40**(S1):S67-S91

[17] Wilbert J, Simoneau K. Folk literature of the Gê Indians. In: Wilbert J, editor. UCLA Latin American Studies. Vol. 44. Los Angeles: UCLA Latin American Center Publications; 1987

[18] Fariña RA, Vizcaíno SF, De Iuliis G. Megafauna: Giant Beasts of Pleistocene South America. Bloomington: Indiana University Press; 2013

Chapter 3

Position of the Ceremony with the Psychedelic Drink Ayahuasca in the Society of the Amero-Indians of the Amazon

Matea Stiperski Matoc, Zoran Stiperski and Tomica Hruška

Abstract

The Amero-Indians of the Amazon traditionally use ayahuasca for various physical and psychological ailments. Shamans in the Peruvian Amazon use the psychedelic drug ayahuasca to have various experiences in guided rituals. The goal of this religious experience is to expand consciousness and gain deeper insights. Consumption of the drink can cause harmful psychotic and paranoid reactions, but the therapeutic value is paramount, namely, the expansion of consciousness, the sensitization to psychological problems, and the search for their solution. The experience gained in the ayahuasca ritual is colorful yet indescribable, so painting pictures serves to better understand the problem that the person is trying to solve. After the ayahuasca ritual, the Amazon shamans knew how to express their expanded consciousness and deep knowledge through painting. The paintings that emerge after the ayahuasca ceremony are imbued with their cosmology of understanding the world.

Keywords: Peruvian Amazon, indigenous people, ayahuasca, traditional healing, shaman ritual, shaman painting, westernization of ayahuasca

1. Introduction

We have tried to explain the complex subject of ayahuasca from various perspectives: geography, sociology, ethnology, history, pharmacology, psychiatry, and art history. The team of authors consists of a doctor (Matea Stiperski Matoc), a geographer (Zoran Stiperski), and a priest (Tomica Hruška). The photos show communities in the Peruvian Amazon region at the time when the Franciscan Tomica Hruška was working as a missionary (1999–2006) and Matea Stiperski Matoc and Zoran Stiperski were traveling (2011).

The experiences and knowledge of Franciscan Tomica Hruško were important for understanding the topic of ayahuasca. The Franciscan Tomica Hruška worked for 8 years as a missionary in the Peruvian Amazon region together with the Franciscan Anton Gerard Žerdin. The Croatian Franciscan Gerard (1950-) has been working as a missionary in the Peruvian Amazon region since 1975 (he was also a missionary during the time of the Sendero Luminoso). Gerard has been bishop in the Peruvian

Amazon region since 2002. The diocese, based in San Ramon, covers a part of the Peruvian Amazon that is as large as Austria. Bishop Gerard founded the Catholic College of Nopoki in Atalaya as a branch of the Universidad Catolica Sedes Sapientiae in Lima with the aim of training the indigenous population to become teachers in schools and evangelization assistants in indigenous languages. Interestingly, they learn various trades that are useful to the indigenous communities, such as building houses and furniture, making clothes and basic agricultural activities. The basic idea of the Nopoki College is to educate and train the indigenous people of Amazonia to preserve their customs, roots, and languages. Nopoki College is committed to an ecologically sustainable economy in the Peruvian Amazon region. Bishop Gerard's many years of knowledge of the Peruvian Amazon region were important for the missionary Tomica. It is important to emphasize that the missionaries are connected to the local community and very familiar with the social reality.

Matea and Zoran traveled to the Peruvian Amazon region together with Bishop Gerard in 2011. Incidentally, you do not travel to this part of Peru with tour operators, but with locals, or in our case with the bishop and his missionaries. The local, indigenous population is afraid of foreigners and strangers. The impression of remoteness and ignorance of the Peruvian Amazon region were best illustrated when we were asked about the Amazon region by Peruvians from Lima after our trip to the Peruvian Amazon region in 2011. It turned out that we, the foreigners, were explaining life in Peru to the Peruvians.

The world sees the Amazon as a vast, distant, unknown rainforest inhabited by mysterious peoples. The Amazon is a tropical rainforest with the highest biodiversity. It is the largest oxygen producer on earth. By preserving the autochthony of the Amazon peoples, their knowledge is preserved from disappearing [1].

Ayahuasca is a sacred drink used by many indigenous peoples of the Americas in the Amazon [2]. It is traditionally used in a special shamanic healing ceremony [3]. The name ayahuasca is derived from the Quechuan words *aya* (spirit or dead) and *huasca* (vine or rope), meaning 'vine of spirits' or 'vine of the dead.' Ayahuasca has several names in the Amazon, for example, the Tukano name *yagé* or *yajé* is used in Colombia, the Shuar term *natém* in Ecuador, and *Daime* and *hoasca* in Brazil [4]. The National Institute of Culture of Peru declared the use of ayahuasca as a Peruvian national heritage in 2008, when it was recognized as part of the traditional medicine of the indigenous peoples of the Amazon [5].

Ayahuasca has been studied by anthropologists, ethnobotanists, neurophysiologists, and psychiatrists [6]. Anthropology talks about the important role of ayahuasca consumption through structured rituals for the indigenous peoples of the Amazon, highlighting their intertwining of religious rituals, belief systems, cosmovisions, artistic productions, music, and healing practices [7]. Ayahuasca becomes a means of shamanic quest and a deity in the pantheon amidst the wilderness [8]. Shamans who lead the traditional ayahuasca ceremony are called *curanderos*, *vegetalismo*, or *ayahuascero*. They must undergo an initiation rite to be allowed to perform the ceremony [9]. The knowledge of how to perform the ceremony is passed down through oral tradition. Besides the psychedelic drink ayahuasca, the ceremony consists of environment, dance, and story told by the shaman. The goal of the rituals is to heal people mentally and physically. The participants in the ceremony believe that chanting creates the inner conditions for healing [10]. Central to healing is the psychological connection with nature [11, 12]. Traditional societies see themselves as part of nature. In the shamanic world, 'nature,' 'culture,' and 'health' are one and the same and inseparable.

With the expansion of Western civilization began the arrival of doctors and the construction of hospitals in the Amazon. The need for indigenous shamanic ayahuasca ceremonies to treat the diseases of the Amazon population diminished. The population is turning to Western medicine, especially the younger people. However, older people, who are more suspicious of modern medicine and do not speak Spanish, still use shamanic ceremonies for healing purposes. On the other hand, younger shamans are increasingly performing ayahuasca ceremonies for tourists, and the ceremony itself is gradually changing and losing its authenticity. Recently, shamanic ayahuasca ceremonies have become more popular in modern Western society [13]. People in the West use them for the treatment of mental disorders and drug addiction [14], as well as for spiritual development and psychological therapy.

2. Anthropology of ayahuasca in the indigenous Amazon

The area of traditional use of the psychedelic drug ayahuasca and the shamanic ayahuasca ritual is associated with the indigenous communities of the Amazon. Similar psychedelic drugs have also been used in a similar manner in other traditional communities outside the Amazon. The long history of altering consciousness through the ritual use of ayahuasca dates back to 2000 B.C. in Ecuador and as far back as 3000 B.C. in the upper Amazon [15]. Material evidence of the use of plant hallucinogens in the Ecuadorian Amazon 1500–2000 B.C. includes ceramic dishes, snuff trays, and pipes [16]. Some authors believe that cave paintings made thousands of years ago in different parts of the world are the result of rituals similar to Ayahuasca rituals [17]. Spanish and Portuguese explorers first learned about ayahuasca when they arrived in the Amazon in the early sixteenth century [18]. In 1616, the Holy Inquisition condemned the ceremonial use of hallucinogenic drugs, believing the effects to be the work of the devil [19]. The first descriptions of the substances that are components of ayahuasca come from Christian missionaries from the nineteenth century [20]. Ethnobotanist Richard Spruce discovered *Banisteriopsis caapi* during his expeditions through the Amazon and the Andes between 1849 and 1864 when he researched the components of ayahuasca [21]. Richard Spruce saw indigenous Tukanos drinking ayahuasca in Brazil in 1851 [7]. The first report of ayahuasca use on the Rio Napo in Ecuador was by geographer Manuel Villavicencio in 1858 [22]. Early on, it was learned that the plant was used for 'journeys to other worlds' and 'visits to tribal gods' [23]. In the twentieth century, ayahuasca was recognized in Brazil as a sacred beverage for syncretic Christian, Spiritist, and African-descended religions: Santo Daime, União do Vegetal, and Barquinha [22, 24–27].

The ayahuasca ritual is deeply rooted in the indigenous communities of Amazonia (**Figures 1** and **2**). Archeological research has shown that the prehistoric Amazon was rarely inhabited by hunters, fishermen, and fruit gatherers due to harsh natural conditions and poor soils [28–32]. Rare small settlements are scattered throughout the vast forests. There are no material remains of abandoned cities and highly developed civilizations in the Amazon. The traditional inhabitants of the Amazon live off the forest and water. They hunt, fish, and gather fruits, plants, honey, insects, fish, crustaceans, lizards, and reptiles [33]. Hunting-fishing-gathering-basic agricultural activities allow for a very low population density in the Amazon [1, 33]. The forests and rivers provide habitat for only a few hunters, fishers, fruit gatherers, and traditional farmers [34]. All indigenous settlements in Amazonia have very small populations (**Figures 3** and **4**), often only extended families or relatives live in

Figure 1.
The Ashaninka group along the Ucayali River. Photo taken in 1999. Picture by the author Tomica Hruška.

Figure 2.
The indigenous population of the Peruvian Amazon. Photo taken in 1999. Picture by the author Tomica Hruška.

them, sometimes at most a few families [1]. From population history studies, we can conclude that larger rivers were the main magnet for sedentary populations because of the resulting fertile soil [35–37] and abundance of fish [37, 38].

Figure 3.
The village of Buenos Aires along the Tambo River. Photo taken in 2011. Picture by the author Zoran Stiperski.

Figure 4.
Catechesis of Ameroindian children under the canopy in the Peruvian Amazon. Photo taken in 1999. Picture by the author Tomica Hruška.

Societies have their own methods of treating disease. The indigenous peoples of the Amazon use the ayahuasca ritual to treat various physical and psychological causes of disease [39]. Medicine and religion were not separate in the Amazonian shamanic age, as shown by the term 'shamanic complex,' which is a 'healing mechanism' [40]. The Tukano (Colombia and Brazil), the Shipibo-Conibo (Peru), and the Shuar (Ecuador), as well as the 72 other tribes, use the ayahuasca drink as a pharmacopeia and care system [41]. However, it must be emphasized that ayahuasca is only one of many plants in the complex traditional pharmacopeia of Amazonia [42, 43]. The ritual ayahuasca treatment in indigenous communities has similarities with magic and witchcraft [41].

3. Pharmacology of ayahuasca

Ayahuasca is a psychoactive herbal tea prepared by boiling the bark and stems of the *Banisteriopsis caapi* vine and the leaves of the *Psychotria viridis* shrub. *B. caapi* contains β-carboline, harmine, harmaline, and tetrahydroharmine (THH), whereas *P. viridis* contains dimethyltryptamine (DMT) [44, 45]. The main component of the tea is the serotonergic, psychedelic, but orally labile DMT. Monoamine oxidases (MAO) in the intestine and liver break down DMT and prevent its absorption into the bloodstream. β-Carbolines are potent reversible and competitive inhibitors of peripheral MAOs that prevent peripheral degradation of DMT in the gastrointestinal tract, allowing them to reach the central nervous system [46]. In addition, they block the deamination of serotonin, increasing its levels in the brain [47]. In the central nervous system, DMT binds to the serotonin receptors 5-HT1 and 5-HT2A, where it acts as a partial agonist [48]. In addition, there is evidence that THH, one of the beta-carbolines, acts as a selective serotonin reuptake inhibitor [49, 50]. Inhibition of both systems, MAO and serotonin reuptake, results in increased levels of serotonin in the brain [49, 51].

Ingestion of ayahuasca has been shown to activate serotonin receptors (5-HT 1A/2A/2C) in the paralimbic brain structures that process emotions and self-perception of bodily processes, that is, interoception [46]. Functional magnetic resonance imaging (fMR) showed increased activity in the occipital, temporal, and frontal lobes involved in vision, memory, and intention [52]. Previously, ayahuasca-induced visual stimuli were thought to be generated by associative visual areas rather than the primary visual cortex in the occipital lobe (Brodmann area 17, BA17) [53]. Nonprimary visual areas are activated during psychopathological hallucinations and physiologically during normal dreams in the REM phase [54–56]. In recent studies, increased activity of BA17 was observed after ayahuasca intake, suggesting that ayahuasca-evoked visions may be triggered in the primary visual cortex. The measured activity of BA17 is highest when subjects look with their eyes open, regardless of ayahuasca ingestion. When eyes are closed, BA17 activity is very low before ayahuasca ingestion, whereas it is high after ingestion (**Figures 5** and **6**). The functional prevalence of BA17 after ayahuasca consumption suggests that the visions, which are strong even with eyes closed, originate in the primary visual cortex. In addition to the visual cortex, ayahuasca enhanced the activity of Brodmann cortical areas 30 and 37, which are involved in the retrieval of episodic memory and the processing of contextual associations. Cortical regions necessary for the integration of individual visual elements into a whole are activated. Activity in Brodmann area 10, a frontal area involved in intentional prospective imagination, working memory, and processing information

Figure 5.
Ashaninkas in national costume. Photo taken in 2004. Picture by the author Tomica Hruška.

Figure 6.
Traditional face decoration with the natural color of the Echota plant. Photo taken in 1999. Picture by the author Tomica Hruška.

from internal sources, is also increased [52]. Frontal, temporal, and cingulate gyrus also increased during performance [57].

After oral ingestion of ayahuasca, it takes 30–40 minutes for the effects to set in. There is an increase in blood pressure, an increase in the number of heartbeats and breathing rates, an increase in body temperature, and dilation of the pupils. The duration of the effect is up to 4 hours [45]. Psychological effects include emotional, sensory, and cognitive changes, as well as visual hallucinations [52]. Visual hallucinations are referred to as visions or *mirações* [58, 59]. Visions occur in a variety of contexts, from simple to complex situations. They are often described as a dream in intense colors. They may see a particular animal or even talk to strangers. All visions of ayahuasca occur from within, that is, without an external stimulus [59].

Studies that have investigated the safety of ayahuasca have proven that it is not harmful to health [26, 45, 58, 60]. Changes such as the increase in heart rate and blood pressure were not significant [45]. It has also been proven that the ritual use of tea is not associated with psychosocial problems that occur with the abuse of other psychoactive drugs [61].

4. Healing through the ayahuasca ritual

Indigenous cultures use various herbal preparations to connect with nature and communicate with spirits beyond our world [3]. Psychedelic plants and mushrooms are used in various forms of religious and cultural life [15, 62, 63], such as initiations into adulthood, seasonal rites, gatherings, and preparations for war [64]. Amazonian shamans use the psychedelic drug ayahuasca to gain different experiences in guided rituals. The goals of this partially religious ritual are the expansion of existing consciousness, the discovery of new states of consciousness, and the attainment of deeper knowledge. During the ceremony, ayahuasca heightens the senses and induces hallucinations based on the participants' subconscious. Ameroindians in the Amazon traditionally use ayahuasca as a medicine for various physical and psychological ailments. In the absence of doctors, Ameroindians sought help from *curanderos* for their health problems. Healing is the main purpose of ayahuasca consumption.

Understanding the effects of the ayahuasca ceremony requires an integrative and multidisciplinary approach. Central to this approach is the close connection with nature. According to indigenous shamanic culture, the healing process is linked to the existing relationship between humans and nature [65]. The essence of spirituality and the view of life of traditional indigenous communities is that they do not control nothing, do not own nothing, and that all nature is given to them to use, and they take only as much as they need. In the ritual, the boundaries between the patient, the doctor, the drug, and the environment are removed. During the ceremony, self-awareness disappears, and one becomes one with the environment and the universe. A person's mental and physical health, as well as the health of the community and the environment, are considered whole rather than separate entities [66]. People react differently during the ceremony depending on the different physical reactions to the active ingredients of ayahuasca, as well as different personal experiences, personalities, environments, and the shaman himself. The shaman reacts depending on the reactions and behavior of the participants in the ceremony. The important role of the shaman in the ritual is to keep the participants of the ritual sane and to ensure that their consciousness does not 'wander off.'

Ceremony participants are expected not to consume meat but to eat only natural and unprocessed foods for several weeks before and after the ceremony [11]. Shamans believe that an unhealthy diet contributes to evil spirits attacking a person during a ceremony. A plant-based diet is consistent with ayahuasca and the rainforest [11]. Purification in the form of vomiting or sweating is an important moment in the ceremony. Purification removes negative emotions, that is, it is proof of a successful healing.

The healing of participants in the ayahuasca ceremony is achieved not only with the help of the drink but also through special songs, dances, and the whole ritual that connects the participant with the community and nature [12]. *Ayahuasceros* use sacred chants to invite the spirits to try to heal the participants of the ceremony [67]. The shaman uses sounds through music (chanting, percussion, whistling) to evoke visual stimuli in participants. Many participants were able to indicate how they 'saw music' [10]. The rapid change of intense colors, shapes, and geometric patterns triggered by the ayahuasca ceremony becomes a unique, never-before-experienced experience for the individual [67]. The shamans' ceremonial chants are adapted to the current situation, environment, and participants [68]. Shamans believe that a person's physical health improves by improving their mental health and vice versa. According to traditional beliefs, a sick person is possessed by a demon or an evil spirit, and the evil forces are expelled from the body through a spiritual ritual [65]. Hallucinogenic drugs evoke visions of spirits manifesting in material forms such as jaguars, birds, and snakes [69]. The healing and victory over diseases of participants in the ceremony may be represented by visions of a snake shedding its skin [65]. The Shipibo believe that they travel to the underworld through hallucinations and encounter a spirit that threatens a sick person, and that the shaman invites good and great spirits to help with healing through chants [70]. The patient has the feeling that during the ritual, he is in a special environment in a kind of sacred space.

Amazonian peoples considered ayahuasca as a carrier of messages from the other world and served as a source of knowledge about other worlds [71]. Ayahuasca and other psychoactive drugs were the only path to true knowledge for many Native American communities [72, 73]. Visions based on personal experiences have special meaning because they convey messages [71]. Rituals bring our hidden memories and feelings back to light and connect us to our deepest feelings and thoughts [74]. The whole ceremony evokes strong religious and spiritual feelings. For many participants in ayahuasca ceremonies, the visions have been radically transformative. There have been cases where people had such a shocking experience during the ritual that they did not want to talk about it, but after the ritual, they became 'better' people.

The therapeutic effects of ayahuasca must be considered within a bio-psycho-socio-spiritual framework [75]. The historical and indigenous context is particularly important for understanding ayahuasca. The environment of the Amazon forest and the community is part of the healing. Shamanic rituals are difficult to perform in modern psychiatric institutions. Healing through an ayahuasca ceremony cannot be compared to simply taking medication, as it has a psychosocial and spiritual component in addition to the pharmacological (**Figures** 7 and **8**). Psychiatry has developed a biopsychosocial model that views illness and health as an interrelationship of biological, psychological, and social states [76]. The integrated mental health model lacks the connection to community, nature, and spirituality that is the foundation of shamanic rituals [77]. There are considerations that the exclusively rational and scientific view in psychiatry is too limited [78]. Western scientists strive to understand the healing effects of psychedelic substances, whereas shamans view medicine, environment, and ceremony as an inseparable complex.

In the early twentieth century, the popular Zen philosopher D. T. Suzuki encouraged military self-sacrifice through the mystical dissolution of the ego [79]. The fragmentation of the ego can be used for violent political purposes. During the secret operation MK-ULTRA in the 1970s, the U.S. military used LSD to militarize the mind

Figure 7.
Children under the canopy. Photo taken in 1999. Picture by the author Tomica Hruška.

Figure 8.
Children on the terrace in front of the house. Photo taken in 2011. Picture by the author Matea Stiperski Matoc.

DOI: http://dx.doi.org/10.5772/intechopen.1003821

by destroying the ego [80]. Psychedelic drugs produced axiosity and crushed people through psychological pressure [81]. The Shuar warriors in Ecuador and Peru used ayahuasca to fight a terrible and powerful spirit that developed murderous abilities [8]. Ayahuasca was used by some indigenous peoples of the Amazon as a bloodthirsty agent in wars. During the rites of aggressive shamanism, people under the influence of ayahuasca knew how to become possessed by ruthless rage against the enemy. The dark side of psychedelics that can dissolve the ego is a warning to all who are completely uncritical of psychedelic alterations of consciousness. The purpose of this mention is not to diminish the positive effects of the substances but to introduce certain negativity, doubt, and caution into the discussion of psychedelics as drugs.

4.1 Experiences with ayahuasca rituals of the Franciscan missionary Tomica Hruška

The ritual ayahuasca experiences of the missionary and Franciscan Tomica Hruška took place in 1999 and 2006 in Atalaya (**Figure 9**), the town where the Tambo and Urubamba rivers merge into the Ucayali River (**Figure 10**) in the Peruvian Amazon. At that time, and especially in 1999, Atalaya was a small settlement that developed into a real port city in the following decade. During that time, the Franciscan Tomica Hruška worked as a missionary in remote parts of the Peruvian Amazon and rarely visited the mission station in Atalaya. The ayahuasca ceremonies were performed by a shaman from Atalaya in the traditional manner, although the authenticity of the ayahuasca ritual, in general, is threatened by the commercialization of ayahuasca for

Figure 9.
The city of Atalaya. Photo taken in 2011. Picture by the author Zoran Stiperski.

Figure 10.
The pier of Atalaya on the river Ucayali. Photo taken in 2011. Picture by the author Zoran Stiperski.

tourism. The photographs in this chapter were taken in a larger area along the Ucayali River and are intended to introduce us to the Amazonian communities where the use of ayahuasca was common at the turn of the twenty and twenty-first centuries.

The first ritual, 1999. The liana (*Banisteriopsis caapi*) was chopped into smaller pieces and boiled with other ingredients for hours until a thick, resinous mixture was formed that was then drunk diluted with water. The preparation of ayahuasca itself is ritualistic and spiritual. During the preparation of ayahuasca, which lasted up to 2 hours, we had conversations. The ritual took place in the afternoon, after the preparation of the drink. The ritual took place in a leafy courtyard filled with peace. Since Tomica was tasting ayahuasca for the first time, the curandero gave him a weaker ayahuasca. Soon, he felt a slight pain in the back of his head. He was conscious the whole time. He heard the *curanderos* reciting songs and asking spiritual beings for help, but he also watched over Tomica to keep him calm. Tomica became aware of his body. He looked at himself in a positive way. His body appeared crystal clear. He saw the light with his eyes closed. When he opened his eyes, he saw a normal, ordinary environment. When he closed his eyes again, a series of strong visions continued, which he describes as sparkling, rich abstractions and various unknown colors. A very unreal and beautiful experience. Tomica was happy and unencumbered in everyday life afterwards. The visions lasted for about 2 hours. The only consequence is the expected purification of the organism.

Second ritual, 2006. After 7 years, the second ayahuasca ceremony took place. At that time, Tomica was very burdened in his daily life. Since the city of Atalaya had expanded in recent years, the courtyard where the last ceremony took place was no longer nice and quiet on the outskirts of the city, but in the middle of the city, surrounded by the noise of people, cars, motorcycles, and tuk-tuks. The ceremony of Tomica was diametrically different from the first ritual in 1999. There was no more positive light. All energy no longer emanates from the entire body. The second experience was the opposite, he describes it as a helmet on the head with only one hole. Fear and discomfort dominated. Tomica cannot remember much about the ayahuasca ceremony in question. *Curanderos* noticed this discomfort.

The conclusion of the Franciscan Tomica is that the visions and experiences of the rituals are a reflection of our personal state or even an amplifier of our current inner state. Tomica felt good in everyday life and therefore had a positive vision in 1999, whereas 2006 was a dark time in his life, and the ayahuasca vision was filled with discomfort and anxiety.

5. Artistic record of the ritual experience

Through visionary art, the nebulous becomes visible and conscious. Visionary artists depict visions evoked by psychoactive drugs. Anthropologists cite the Amazonian plant *Banisteriopsis caapi*, from which ayahuasca is derived, the mushroom *Amanita muscaria* used by ancient Hindu shamans, and the peyote cactus used by Native Americans, as traditional sources of entheogenic substances. Contemporary visionary artists speak of synthetic psychoactive substances such as LSD as a means of achieving higher states of consciousness [17].

The experiences and visions one has during an ayahuasca ceremony are almost indescribable. Amazon shamans use painting to express their expanded consciousness and deep knowledge after the ayahuasca ritual. The paintings are steeped in the traditional cosmology of understanding the world and the personal experiences on which the visions are based (**Figures 11** and **12**). In a traditional ayahuasca ceremony, the *curanderos* drink the potion with the patient and have visions that are responses to the diseases and disorders that plague the patient. On the basis of the painted picture, through which he expresses his visions, the shaman diagnoses the disease. The visions consisted of vivid colors, not necessarily regular geometric abstractions, but a kind of natural geometry in the sense that people and nature were geometrized. The visual experiences ranged from abstract shapes such as circles, triangles, and spirals to culturally specific images such as jaguars, snakes, and mythical landscapes [82]. Animal forms frequently appear in shamanic ceremonies of various cultures around the world and are common motifs in ancient rock paintings [83]. Ayahuasca vision is considered a state of consciousness superior to the normal state of consciousness [3].

The creativity of the Shipibo people in the Peruvian Amazon is evident in the design of useful items and paintings inspired by ritual ayahuasca visions [70]. Patterns on ceramics, clothing, painted bodies, and houses were part of visions during rituals or were dreamed [84]. The vivid visual hallucinations included both intricate geometric shapes and realistic scenes. They considered visions as medicine and painted them on faces, pots, weapons, and houses to protect them. Drawings on the faces of children protected them from disease and death [70]. Patterns inspired by ayahuasca are also found in art, architecture, and as decorations on Tukano pottery and musical instruments [21].

Unusual experiences created by psychedelics are the source of art of our ancient ancestors [85]. The cave drawings at Chauvet (France) and among the San in the

Figure 11.
Ashaninka mother with two children. Photo taken in 1999. Picture by the author Tomica Hruška.

Figure 12.
Children in the Peruvian Amazon. Photo taken in 2011. Picture by the author Matea Stiperski Matoc.

Drakensberg Mountains of South Africa are considered the work of ancient visionary artists. The motifs, otherworldly beings, and stylistic expressions of the ancient cave artists are similar to those of contemporary artists of shamanic visions [86]. Common to all visionary arts are altered states of consciousness, shamanism, and the immediate experience of unusual worlds [86].

6. The role of ayahuasca in modern medicine

In recent decades, the effects of ayahuasca in the treatment of depression and anxiety have been studied [75, 87–90]. In a study of patients suffering from depressive disorder, symptoms and signs of depression decreased immediately after ingestion of ayahuasca, and the effect lasted for 14 days. It is interesting to note that 14 days was the interval between shamanic rituals in Brazil. More importantly, the effect of ayahuasca began much earlier than conventional antidepressants, as early as 40 minutes after ingestion of the tea [91]. Ayahuasca has also been shown not to cause significant sensory, cognitive, or affective changes at lower doses sufficient to reduce symptoms of depression. It is very important that ayahuasca does not induce mania in patients with mood disorders [92]. Studies have shown that regular ayahuasca consumption can modulate the serotonergic system in the brain in the long term, that is, increases the levels of serotonin transporter (SERT), which explains its beneficial effects in the treatment of depression [91]. A reduction in feelings of hopelessness and panic has been noted in members of the Santo Daime Church who have been taking ayahuasca regularly for at least 10 years [93]. In addition, after consuming ayahuasca, people with social anxiety have more self-confidence when speaking in public [94]. Long-term ayahuasca intake had no negative effects on personality, psychopathology, neuropsychology, outlook on life, or psychosocial well-being; on the contrary, ayahuasca users showed fewer psychopathological symptoms, better scores on neuropsychological tests, higher levels of spirituality, and better psychosocial adjustment than the control group [95].

In addition to its anxiolytic and antidepressant effects, ayahuasca is considered effective in the treatment of addiction [18, 47, 61, 75]. Treated patients reported therapeutic effects such as increased body awareness, decreased craving for drugs, stimulated emotional processes (catharsis, awareness of previously suppressed emotions), self-analysis, and increased self-efficacy [96]. Furthermore, in most studies, fewer religious ayahuasca users had alcohol problems than the control groups [60, 61]. Interestingly, most people had a history of alcoholism, drug use, and domestic violence before joining the União do Vegetal Church in Brazil. Through the religion and regular intake of ayahuasca, their addiction and dysfunctional behavior disappeared [60]. In 1992, the Takiwasi Center for the Rehabilitation of Drug Addicts and Research of Traditional Medicines was founded in Peru. The centre uses a mixed approach of Western medicine and traditional Amazonian medicine based on local herbs, including ayahuasca [75]. Although this form of treatment has been known in South America for many years, more and more Americans and Europeans are becoming interested in ayahuasca and shamanistic healing, which is why they travel to the Amazon to participate in traditional ceremonies [97].

Finally, it is important to mention the effects of ayahuasca on the endocrine and immune systems. Two hours after ingestion of ayahuasca, an increase in prolactin and cortisol was measured, which returned to their original level within 24 hours. Increased cortisol decreases the proportion of CD3 and CD4 lymphocytes and increases the proportion of natural killer (NK) cells. This effect is transient and returns to baseline levels after 24 hours. A decrease in CD3 and CD4 lymphocytes is considered harmful, whereas an increase in NK cells is beneficial because they are an important component of immunity against virus-infected cells and cancer cells.

Acute stress also leads to an increase in cortisol levels and a redistribution of lymphocytes. According to recent research, acute stress is thought to have a modulatory rather than an inhibitory effect on the immune system. Ayahuasca showed significant neuroendocrine stimulation and modulatory effect on cell-mediated immunity [98]. Interestingly, shamans in South America lived to a ripe old age in better physical and mental health than most Westerners. This is due in part not only to a healthy diet but also to the regular intake of ayahuasca (often several times a week) throughout most of their lives [41, 99].

7. The spread of ayahuasca from the traditional Amazon world

Ayahuasca has spread from the traditional world of the Amazon to numerous countries in South and North America, Europe, Australia, New Zealand, and some Asian countries [100]. Various experiences with ayahuasca, accompanied by the media and the Internet, attract various recreational and professional circles [100]. The syncretic religious group of the Church of Santo Daime in Brazil uses ayahuasca as a sacrament. They consider ayahuasca and other psychedelic drugs as the only path to true knowledge [71]. The Brazilian União do Vegetal is a Christian spiritualist religion present in Spain and the United States that uses psychoactive tea from the Amazon rainforest in its rituals [5]. The use of the ayahuasca beverage in religious ceremonies is officially recognized and legally permitted in Brazil for a few religious groups [101]. It is estimated that there are at least 15,000 urban monthly ayahuasca users in South America [26]. In particular, the churches of the Brazilian syncretic ayahuasca religion pose a major challenge to Western liberal democracies that advocate criminal drug control and constitutionally guaranteed religious freedom [100]. There have already been court cases in Australia, France, Germany, Italy, the Netherlands, Spain, and the United States over the religious use of ayahuasca [100, 102].

In Australia, rituals are performed by neo-shamans using native acacia trees [103]. Traditional shamanistic ceremonies are changing due to colonialism and globalization [104]. With the globalization of cultures, the roots of the geographical, social, and cultural context of the ayahuasca ritual disappear [100]. Shamanic ayahuasca tourism is spreading worldwide as shamanic ayahuasca ceremonies have become increasingly popular in recent decades (**Figures 13** and **14**). Four main reasons for shamanic tourism in the Peruvian Amazon are cited as follows: self-exploration and spiritual growth; curiosity; physical and emotional healing; and vacation in an exotic place [105]. Authenticity, the archaic past, a sense of spirituality, and sacredness are important to tourists [105]. The feeling of emptiness in one's own culture and the absence of social traditions encourage the desire to experience shamanism [9, 14]. The spread of ayahuasca ceremonies outside the Amazon raises the question of cultural appropriation of traditional indigenous knowledge [100]. Neo-shamans and especially false shamans often dilute the ancient, systematized tradition before the eyes of tourists and the contemporary non-Amazonian population in the Amazon. The commercialization of ayahuasca erases ancient knowledge and adapts to new concepts of spirituality and culture. The benefits and harms of ayahuasca can help inform strategies for responsible and effective regulation of its use. Ayahuasca could gain the status of a traditional indigenous medicine and sacrament in religious communities [100].

Figure 13.
Children of the Ashaninka. Photo taken in 1999. Picture by the author Tomica Hruška.

Figure 14.
Children of the Ashaninka. Photo taken in 2011. Picture by the author Zoran Stiperski.

8. Conclusion

The main objective of this chapter is to clarify the meaning of the psychedelic drug ayahuasca and consequently of the shamanic ayahuasca ceremony among the indigenous peoples of Amazonia. In accordance with the constant changes in the world, there is a transformation of the use of ayahuasca in the Amazon and its emergence in modern Western societies. 'Travels' to 'other worlds' with psychedelic drugs have been known for a long time. The first records of the ritual use of ayahuasca date back to 3000 B.C. in the upper Amazon. Spanish and Portuguese explorers of the Amazon introduced Europeans to ayahuasca in the early sixteenth century. Ayahuasca is obtained as a psychoactive herbal tea by long boiling of the bark and stems of the *Banisteriopsis caapi* vine and the leaves of the *Psychotria viridis* shrub.

The traditional Amazon population engaged in hunting, fishing, fruit gathering, and rudimentary agriculture. Hunter-gatherer and slash-and-burn agriculture do not allow dense settlement in the inhospitable environment of the Amazon rainforest. Long-term isolation from higher civilizations has been their historical reality. The absence of any form of modern medicine based on Western principles of evidence required them to develop specific creative responses, as societies tend to develop models to cure disease. The main goal of the ayahuasca ceremony is to diagnose the disease and heal the person. Thus, a shaman is first and foremost a doctor in the modern sense of the word. Because of the traditional intertwining of medicine and religion, he is a shaman and a priest. The psychedelic drug ayahuasca was a means of diagnosing diseases and medicine itself. Ayahuasca was taken in the form of a systematized ritual under the guidance of a shaman. Participants in ayahuasca ceremonies experience various visions that are intense and difficult to describe. As visual artists, the shamans attempt to materialize and describe their visions by painting pictures. The painted pictures become a diagnosis of the disease.

The intrusion of Western influences into the Amazon through Christian missionaries, oil companies, farmers, planters, and armies, as well as the efforts of national governments to implement numerous development projects in energy, communications, agriculture, and industry, and the immigration of population are transforming traditional Amazonian societies. With the advent of hospitals and modern doctors, the importance of shamans and ayahuasca ceremonies is diminishing, especially in the treatment of physical ailments. Shamanic ayahuasca rituals in the Amazon have attracted tourists for several decades. Shamans and neo-shamans are facing new challenges, but new business opportunities are also opening for them. Westerners primarily hope for new insights and experiences, self-exploration, and emotional healing from shamans. The ayahuasca ritual faces transformational challenges. Brazil's new syncretic Christian, Spiritist, and Afro religions, which use ayahuasca for their religious rituals, have put national governments throughout the liberal world in a quandary over how to balance constitutionally guaranteed religious freedoms with criminal control of psychedelic drugs. In the West, there are numerous court cases about the procedure for the permitted use of ayahuasca. In 2008, Peru recognized the use of ayahuasca as traditional medicine and Peruvian national heritage. Ayahuasca is attractive to many people in North and South America, Europe, Australia, New Zealand, and some Asian countries. This attraction is supported by recent research showing that supervised ingestion of ayahuasca has a positive effect in treating depression and anxiety.

Author details

Matea Stiperski Matoc[1]*, Zoran Stiperski[2] and Tomica Hruška[3]

1 Clinical Hospital Dubrava, Zagreb, Croatia

2 Faculty of Mathematical and Natural Sciences, Department of Geography, University of Zagreb, Zagreb, Croatia

3 The Franciscan Monastery of the Holy Spirit, Požega, Croatia

*Address all correspondence to: mstiperski@kbd.hr

References

[1] Stiperski Z, Hruška T, Heršak E, Lorber L. Planimetry of settlements as a confirmation of transformation in the Atalaya region in Peruvian Amazonia. Anthropological Notebooks. 2018;**24**(2):75-96

[2] Labate BC, MacRae E, editors. The Light from the Forest: The Ritual Use of ayahuasca in Brazil. Fieldwork in Religion, 2(3). Equinox: Sheffield; 2006. p. 556

[3] Krippner S, Sulla J. Spiritual content in experiential reports from ayahuasca sessions. NeuroQuantology. 2011;**9**(2):333-350

[4] Apuda I, Romaníd O. Medicine, religion and ayahuasca in Catalonia. Considering ayahuasca networks from a medical anthropology perspective. International Journal of Drug Policy. 2017;**39**:28-36. DOI: 10.1016/j.drugpo.2016.07.011

[5] Labate BC, Goldstein I. Ayahuasca–from dangerous drug to national heritage: An interview with Antonio A. Arantes. International Journal of Transpersonal Studies. 2009;**28**(1):53-64. DOI: 10.24972/ijts.2009.28.1.53

[6] Tupper KW, Labate BC. Ayahuasca, psychedelic studies and health sciences: The politics of knowledge and inquiry into an Amazonian plant brew. Current Drug Abuse Reviews. 2014;**7**:71-80

[7] Schultes RE. The beta-carboline hallucinogens of South America. Journal of Psychoactive Drugs. 1982;**14**(3):205-220. DOI: 10.1080/02791072.1982.10471930

[8] Rubenstein SL. On the importance of visions among the Amazonian Shuar. Current Anthropology. 2012;**53**(1):39-79. DOI: 10.1086/663830

[9] Wasson J. The Ayahuasca Experience: A Phenomenological Study of Tourists in Iquitos, Peru [Thesis]. Lund: Lund University; 2020

[10] Bustos S. The Healing Power of the Icaros: A Phenomenological Study of ayahuasca Experiences [Thesis]. San Francisco: California Institute of Integral Studies; 2008

[11] Gearin AK. An Amazonian Shamanic Brew in Australia: Ayahuasca Healing and Individualism [Thesis]. Brisbane: University of Queensland; 2015

[12] Argento E, Capler R, Thomas G, Lucas P, Tupper KW. Exploring ayahuasca-assisted therapy for addiction: A qualitative analysis of preliminary findings among an indigenous community in Canada. Drug and Alcohol Review. 2019;**38**:781-789. DOI: 10.1111/dar.12985

[13] Tupper KW. The globalization of ayahuasca: Harm reduction or benefit maximization? International Journal of Drug Policy. 2008;**19**(4):297-303. DOI: 10.1016/j.drugpo.2006.11.001

[14] Kavenská V, Simonová H. Ayahuasca tourism: Participants in shamanic rituals and their personality styles, motivation, benefits and risks. Journal of Psychoactive Drugs. 2015;**47**(5):351-359. DOI: 10.1080/02791072.2015.1094590

[15] Furst PT. Hallucinogens and Culture. Novato. CA: Chandler and Sharp; 1976. p. 194

[16] Naranjo P. Elayahuasca en la arqueología ecuatoriana [Ayahuasca in

the Ecuadorian archeology]. América Indígena. 1986;**46**(1):117-127

[17] Colombo DE. Cosmic Expressions and Spiritual Revivals within Visionary Art [Thesis]. San Marcos: Texas State University; 2015

[18] Liester MB, Prickett JI. Hypotheses regarding the mechanisms of ayahuasca in the treatment of addictions. Journal of Psychoactive Drugs. 2012;**44**(3):200-208. DOI: 10.1080/02791072.2012.704590

[19] Grob CS, editor. Hallucinogens: A Reader. New York: Jeremy P. Tarcher, Putnam; 2002. p. 298

[20] Sheldrake M. The 'enigma' of Richard Schultes, Amazonian hallucinogenic plants, and the limits of ethnobotany. Social Studies of Science. 2020;**50**(3):345-376. DOI: 10.1177/0306312720920362

[21] Rudgley R. Essential Substances: A Cultural History of Intoxicants in Society. New York: Kodansha International; 1994. p. 195

[22] Metzner R. Sacred Vine of Spirits: Ayahuasca. Rochester, VT: Inner Traditions; 2006. p. 264

[23] Reichel-Dolmatoff G. The cultural context of an aboriginal hallucinogen: Banisteriopsis Caapi. In: Furst PT, editor. Flesh of the Gods: The Ritual Use of Hallucinogens. New York: Praeger; 1972. pp. 84-113

[24] Labate BC, MacRae E. Ayahuasca, Ritual and Religion in Brazil. London and New York: Routledge; 2010. p. 256

[25] Labate BC, de Rose SI, dos Santos GR. Ayahuasca Religions: A Comprehensive Bibliography & Critical Essays. Santa Cruz, CA: Multidisciplinary Association of Psychedelic Studies; 2008. p. 160

[26] Callaway JC, McKenna DJ, Grob CS, Brito GS, Raymon LP, Poland RE, et al. Pharmacokinetics of hoasca alkaloids in healthy humans. Journal of Ethnopharmacology. 1999;**65**(3):243-256. DOI: 10.1016/S0378-8741(98)00168-8

[27] Talin P, Sanabria E. Ayahuasca's entwined efficacy: An ethnographic study of ritual healing from 'addiction'. International Journal of Drug Policy. 2017;**44**:23-30. DOI: 10.1016/j.drugpo.2017.02.017

[28] Gross D. Protein capture and cultural development in the Amazon Basin. American Anthropologist. 1975;**77**(3):526-549. DOI: 10.1525/aa.1975.77.3.02a00040

[29] Meggers BJ. Environmental limitation on the development of culture. American Anthropologist. 1954;**56**(5):801-824

[30] Meggers BJ. Amazonia: Real or counterfeit paradise? The Review of Archaeology. 1992;**13**(2):25-40

[31] Meggers BJ. Biogeographical approaches to reconstructing the prehistory of Amazonia. Biogeographica. 1994;**70**(3):97-110

[32] Meggers BJ. Prehistoric cultural development in Amazonia: An archaeological perspective. Research and Exploration. 1994;**10**(4):398-421

[33] Stiperski Z, Hruška T. Social changes in the Peruvian Amazon due to foreign influence. In: Mikkola HJ, editor. Ecosystem and Biodiversity of Amazonia. London: IntechOpen; 2021. pp. 195-217. DOI: 10.5772/intechopen.94772

[34] Castillo BH. Indigenous Peoples in Isolation in the Peruvian Amazon. Their Struggle for Survival and Freedom. Copenhagen: IWGIA – International

Work Group for Indigenous Affairs; 2004. p. 245

[35] The LD, Amazon U. Ancient Peoples and Places. Southampton: Thames & Hudson; 1970. p. 256

[36] Denevan WM. A bluff model of riverine settlement in prehistoric Amazonia. Annals of the Association of American Geographers. 1996;**86**(4):654-681. DOI: 10.1111/j.1467-8306.1996.tb01771.x

[37] Lombardo U, Prümers H. Pre-Columbian human occupation patterns in the eastern plains of the llanos de Moxos, Bolivian Amazonia. Journal of Archaeological Science. 2010;**37**:1875-1885. DOI: 10.1016/j.jas.2010.02.011

[38] Carneiro R. The history of ecological interpretations of Amazonia: Does Roosevelt have it right? In: Sponsel L, editor. Indigenous Peoples and the Future of Amazonia: An Ecological Anthropology of an Endangered World. Tucson, AZ: University of Arizona Press; 1995. pp. 45-70

[39] Luna LE. Indigenous and mestizo use of ayahuasca. An overview. In: dos Santos RG, editor. The Ethnopharmacology of Ayahuasca. Kerala: Transworld Research Network; 2011. p. 1-21

[40] Winkelman MJ. Shamanism: A Biopsychosocial Paradigm of Consciousness and Healing. Santa Barbara: ABC-CLIO; 2010. p. 309

[41] Luna LE. Vegetalismo – Shamanism among the Mestizo Population of the Peruvian Amazon. Stockholm: Almqvist & Wiksell International; 1986. p. 202

[42] Bussmann RW, Sharon D. Traditional medicinal plant use in Loja province, southern Ecuador. Journal of Ethnobiology and Ethnomedicine. 2006;**2**(44):1-11. DOI: 10.1186/1746-4269-2-44

[43] Bussmann RW, Sharon D. Traditional medicinal plant use in northern Peru: Tracking two thousand years of healing culture. Journal of Ethnobiology Ethnomedicine. 2006;**2**(47):1-18. DOI: 10.1186/1746-4269-2-47

[44] McKenna DJ, Towers GH, Abbott F. Monoamine oxidase inhibitors in south American hallucinogenic plants: Tryptamine and beta-carboline constituents of ayahuasca. Journal of Ethnopharmacology. 1984;**10**(2):195-223. DOI: 10.1016/0378-8741(84)90003-5

[45] Riba J, Valle M, Urbano G, Yritia M, Morte A, Barbanoj MJ. Human pharmacology of ayahuasca: Subjective and cardiovascular effects, monoamine metabolite excretion, and pharmacokinetics. Journal of Pharmacology and Experimental Therapeutics. 2003;**306**(1):73-83. DOI: 10.1124/jpet.103.049882

[46] Riba J, Romero S, Grasa E, Mena E, Carrió I, Barbanoj MJ. Increased frontal and paralimbic activation following ayahuasca, the pan-Amazonian inebriant. Psychopharmacology. 2006;**186**(1):93-98. DOI: 10.1007/s00213-006-0358-7

[47] McKenna DJ. Clinical investigations of the therapeutic potential of ayahuasca: Rationale and regulatory challenges. Pharmacology & Therapeutics. 2004;**102**(2):111-129. DOI: 10.1016/j.pharmthera.2004.03.002

[48] Rabin RA, Regina M, Doat M, Winter JC. 5-HT2A receptor-stimulated phosphoinositide hydrolysis in the stimulus effects of hallucinogens. Pharmacology Biochemistry and Behavior. 2002;**72**(1-2):29-37. DOI: 10.1016/s0091-3057(01)00720-1

[49] McKenna DJ, Callaway JC, Grob CS. The scientific investigation of ayahuasca: A review of past and current research. The Heffter Review of Psychedelic Research. 1998;**1**:65-77

[50] Frecska E, White KD, Luna LE. Effects of ayahuasca on binocular rivalry with dichoptic stimulus alternation. Psychopharmacology. 2004;**173**:79-87. DOI: 10.1007/s00213-003-1701-x

[51] Luna LE. Narrativas da alteridade: A ayahuasca e o motivo de transformac¸ao em animal. In: Labate BC, Goulart SL, editors. O uso ˜ritual das plantas de poder. Campinas: Mercado de Letras; 2005. pp. 333-354

[52] de Araujo DB, Ribeiro S, Cecchi GA, Carvalho FM, Sanchez TA, Pinto JP, et al. Seeing with the eyes shut: Neural basis of enhanced imagery following ayahuasca ingestion. Human Brain Mapping. 2012;**33**(11):2550-2560. DOI: 10.1002/hbm.21381

[53] D'Esposito M, Detre JA, Aguirre GK, Stallcup M, Alsop DC, Tippet LJ, et al. A functional MRI study of mental image generation. Neuropsychologia. 1997;**35**(5):725-730. DOI: 10.1016/s0028-3932(96)00121-2

[54] Allen P, Larøi F, McGuire PK, Aleman A. The hallucinating brain: A review of structural and functional neuroimaging studies of hallucinations. Neuroscience & Biobehavioral Reviews. 2008;**32**(1):175-191. DOI: 10.1016/j.neubiorev.2007.07.012

[55] Braun AR, Balkin TJ, Wesensten NJ, Gwadry F, Carson RE, Varga M, et al. Dissociated pattern of activity in visual cortices and their projections during human rapid eye movement sleep. Science. 1998;**279**(5347):91-95. DOI: 10.1126/science.279.5347.91

[56] Wehrle R, Kaufmann C, Wetter TC, Holsboer F, Auer DP, Pollmächer T, et al. Functional microstates within human REM sleep: First evidence from fMRI of a thalamocortical network specific for phasic REM periods. European Journal of Neuroscience. 2007;**25**:863-871. DOI: 10.1111/j.1460-9568.2007.05314.x

[57] Prado DA, Pinto J, Crippa J, Santos A, Ribeiro S, Araujo D, et al. Effects of the Amazonian psychoactive plant beverage ayahuasca on prefrontal and limbic regions during a language task: A fMRI study. European Neuropsychopharmacolgy. 2009;**19**:S314-S315. DOI: 10.1016/S0924-977X(09)70469-9

[58] Riba J, Rodríguez-Fornells A, Urbano G, Morte A, Antonijoan R, Montero M, et al. Subjective effects and tolerability of the south American psychoactive beverage ayahuasca in healthy volunteers. Psychopharmacology. 2001;**154**:85-95. DOI: 10.1007/s002130000606

[59] Shanon B. Ayahuasca visualizations a structural typology. Journal of Consciousness Studies. 2002;**9**(2):3-30

[60] Grob CS, McKenna DJ, Callaway JC, Brito GS, Neves ES, Oberlaender G, et al. Human psychopharmacology of hoasca, a plant hallucinogen used in ritual context in Brazil. Journal of Nervous and Mental Disease. 1996;**184**(2):86-94. DOI:10.1097/00005053-199602000-00004

[61] Fábregas JM, González D, Fondevila S, Cutchet M, Fernández X, Barbosa PCR, et al. Assessment of addiction severity among ritual users of ayahuasca. Drug and Alcohol Dependence. 2010;**111**(3):257-261. DOI: 10.1016/j.drugalcdep.2010.03.024

[62] Dobkin de Rios M. Hallucinogens: Cross-Cultural Perspectives. Albuquerque: University of New Mexico Press; 1984. p. 255

[63] Harner MJ, editor. Hallucinogens and Shamanism. New York: Oxford University Press; 1973. p. 224

[64] Gearin AK, Devenot N. Psychedelic medicalization, public discourse, and the morality of ego dissolution. International Journal of Cultural Studies. 2021;**24**(6):917-935. DOI: 10.1177/13678779211019424

[65] Blomster KE. How ayahuasca-ceremonies change perceptions of natural values in participants. In: The Links to Nature and the Role of Shamanic Rituals [Thesis]. Uppsala: Uppsala University; 2021

[66] Salmón E. Kincentric ecology: Indigenous perceptions of the human nature relationship. Ecological Applications. 2000;**10**(5):1327-1332. DOI: 10.1890/1051-0761(2000)010[1327:KEIPOT]2.0.CO;2

[67] Dobkin de Rios M. Amazon Healer: The Life and Times of an Urban Shaman. Bridport, Dorsey, England: Prism Press; 1992. p. 180

[68] Mercante MS. Images Ofhealing: Spontaneous Mental Imagery and Healing Process of the Barquinha, a Brazilian ayahuasca Religious System [Thesis]. San Francisco: Saybrook Graduate School and Research Center; 2006

[69] Gow P. Could Sangama read? The origin of writing among the Piro of eastern Peru. History and Anthropology. 1990;**5**:87-103. DOI: 10.1080/02757206.1990.9960809

[70] Hern WM. Yushin Huemena: Visions of the Spirit world, art, design, medicine and protective spirits in Shipibo ritual. Journal of the Society for the Anthropology of lowland South America. 2016;**14**(1):1-14

[71] Shanon B. The epistemics of ayahuasca visions. Phenomenology and the Cognitive Sciences. 2010;**9**:263-280. DOI: 10.1007/s11097-010-9161-3

[72] Taussig M. Shamanism, Colonialism, and the Wild Man: A Study in Terror and Healing. Chicago and London: The University of Chicago Press; 1987. p. 517

[73] Chaumeil J. Voir, savoir, pouvoir: le chamanisme chez les Yagua du nord-est Péruvien. Paris: Editions de l'Ecole des Hautes Etudes en Sciences Sociales; 1983. p. 326

[74] Uzendoski M. Somatic poetry in Amazonian Ecuador. Anthropology and Humanism. 2008;**33**(1/2):12-29. DOI: 10.1111/j.1548-1409.2008.00002.x

[75] Frecska E, Bokor P, Winkelman M. The therapeutic potentials of ayahuasca: Possible effects against various diseases of civilization. Frontiers in Pharmacology. 2016;**7**(35):1-17. DOI: 10.3389/fphar.2016.00035

[76] Engel GL. The need for a new medical model: A challenge for biomedicine. Science. 1977;**196**(4286):129-136. DOI: 10.1126/science.84746

[77] Labate B, Cavnar C, editors. The Therapeutic Use of Ayahuasca. Heidelberg: Springer; 2013. p. 226. DOI: 10.1007/978-3-642-40426-9

[78] Csordas TJ. Body/Meaning/Healing. New York: Palgrave Macmillan; 2002. p. 321. DOI: 10.1007/978-1-137-08286-2

[79] Gier N. The Origins of Religious Violence: An Asian Perspective. Lanham: Lexington Books; 2014. 324 p

[80] Richert L, Dyck E. Psychedelic crossings: American mental health and LSD in the 1970s. Medical Humanities. 2020;**46**(3):184-191. DOI: 10.1136/medhum-2018-011593

[81] Lee MA, Shlain B. Acid Dreams: The Complete Social History of LSD : The CIA, the Sixties, and beyond. Rev. Evergreen Ed. New York: Grove Weidenfeld; 1992. 345 p

[82] Reichel-Dolmatoff G. Shamanism and the Art of the Eastern Tukanoan Indians. Leiden: E. J. Brill; 1987. p. 25

[83] Eliade M. Shamanism: Archaic Techniques of Ecstasy. London: Routledge & Kegan Paul; 1964. p. 610

[84] Feldman NG. Shipibo-conibo textiles 2010-2018: Artists of the Amazon culturally engaged. In: Textile Society of America Symposium Proceedings; 19-23 September 2018; Vancouver, BC, Canada. Baltimore: Textile Society of America; 2018. DOI: 10.32873/unl.dc.tsasp.0029. Available from: https://digitalcommons.unl.edu/tsaconf/1082

[85] Devereux P. The Long Trip: A Prehistory of Psychedelia. London: Penguin Arkana; 1997. p. 298

[86] Pablo HG. Amaringo: A special state of consciousness. In: Charing HG, Cloudsley P, Amaringo P, editors. The Ayahuasca Visions of Pablo Amaringo. Rochester: Inner Traditions; 2011. pp. 10-11

[87] Sanches RF, de Lima OF, dos Santos RG, Macedo LRH, Maia-de-Oliveira JP, Wichert-Ana L, et al. Antidepressant effects of a single dose of ayahuasca in patients with recurrent depression: A SPECT study. Journal of Clinical Psychopharmacology. 2016;**36**(1):77-81. DOI: 10.1097/JCP.0000000000000436

[88] Galvão ACM, de Almeida RN, Silva EADS, Freire FAM, Palhano-Fontes F, Onias H, et al. Cortisol modulation by ayahuasca in patients with treatment resistant depression and healthy controls. Frontiers in Psychiatry. 2018;**9**(185):1-10. DOI: 10.3389/fpsyt.2018.00185

[89] Nunes AA, Dos Santos RG, Osório FL, Sanches RF, Crippa JAS, Hallak JEC. Effects of ayahuasca and its alkaloids on drug dependence: A systematic literature review of quantitative studies in animals and humans. Journal of Psychoactive Drugs. 2016;**48**(3):195-205. DOI: 10.1080/02791072.2016.1188225

[90] Carbonaro TM, Gatch MB. Neuropharmacology of N,N-dimethyltryptamine. Brain Research Bulletin. 2016;**126**(1):74-88. DOI: 10.1016/j.brainresbull.2016.04.016

[91] Palhano-Fontes F, Alchieri JC, Oliveira JPM, Soares BL, Hallak JEC, Galvao-Coelho N, et al. The therapeutic potentials of ayahuasca in the treatment of depression. In: Labate BC, Cavnar C, editors. The Therapeutic Use of Ayahuasca. Berlin, Heidelberg: Springer; 2013. pp. 23-39. DOI: 10.1007/978-3-642-40426-9_2

[92] Osório Fde L, Sanches RF, Macedo LR, dos Santos RG, Maia-de-Oliveira JP, Wichert-Ana L, et al. Antidepressant effects of a single dose of ayahuasca in patients with recurrent depression: A preliminary report. Brazilian Journal of Psychiatry. 2015;**37**(1):13-20. DOI: 10.1590/1516-4446-2014-1496

[93] Santos RG, Landeira-Fernandez J, Strassman RJ, Motta V, Cruz APM. Effects of ayahuasca on psychometric measures of anxiety, panic-like and hopelessness in Santo Daime members. Journal of Ethnopharmacology. 2007;**112**(3):507-513. DOI: 10.1016/j.jep.2007.04.012

[94] Dos Santos RG, Osório FL, Rocha JM, Rossi GN, Bouso JC, Rodrigues LS, et al.

Ayahuasca improves self-perception of speech performance in subjects with social anxiety disorder: A pilot, proof-of-concept, randomized, placebo-controlled trial. Journal of Clinical Psychopharmacology. 2021;**41**(5):540-550. DOI: 10.1097/JCP.0000000000001428

[95] Bouso JC, González D, Fondevila S, Cutchet M, Fernández X, Ribeiro Barbosa PC, et al. Personality, psychopathology, life attitudes and neuropsychological performance among ritual users of ayahuasca: A longitudinal study. PLoS One. 2012;**7**(8):e42421. DOI: 10.1371/journal.pone.0042421

[96] Loizaga-Velder A, Verres R. Therapeutic effects of ritual ayahuasca use in the treatment of substance dependence – Qualitative results. Journal of Psychoactive Drugs. 2014;**46**(1):63-72. DOI: 10.1080/02791072.2013.873157

[97] Dobkin de Rios M. Drug tourism in the Amazon. Anthropology of Conciousness. 1994;**5**(1):16-19. DOI: 10.1525/ac.1994.5.1.16

[98] Dos Santos RG, Valle M, Bouso JC, Nomdedéu JF, Rodríguez-Espinosa J, McIlhenny EH, et al. Autonomic, neuroendocrine, and immunological effects of ayahuasca: A comparative study with d-amphetamine. Journal of Clinical Psychopharmacology. 2011;**31**(6):717-726. DOI: 10.1097/JCP.0b013e31823607f6

[99] Luna LE. The concept of plants as teachers among four mestizo shamans of Iquitos, northeastern Peru. Journal of Ethnopharmacology. 1984;**11**(2):135-156. DOI: 10.1016/0378-8741(84)90036-9

[100] Tupper KW. Ayahuasca healing beyond the Amazon: The globalization of a traditional indigenous entheogenic practice. Global Networks. 2009;**9**(1):117-136. DOI: 10.1111/j.1471-0374.2009.00245.x

[101] Ott J. Ayahuasca Analogues : Pangean Entheogens. Kennewick: Natural Products; 1994. p. 127

[102] Adelaars A. Court Case in Holland against the Use of ayahuasca by the Dutch Santo Daime Church [Internet]. 2001. Available from: https://www.santodaime.org/site/site-antigo/community/news/2105_holland.htm [Accessed: September 28, 2023]

[103] Rodd R. It's all you! Australian ayahuasca drinking, spiritual development, and immunitary individualism. Critique of Anthropology. 2018;**38**(3):325-345. DOI: 10.1177/0308275X18775818

[104] Bussmann RW. The globalization of traditional medicine in Northern Peru: From shamanism to molecules. Evidence-Based Complementary and Alternative Medicine. 2013;**291903**:1-46. DOI: 10.1155/2013/291903

[105] Fotiou E. On the uneasiness of tourism: Considerations on shamanic tourism in Western Amazonia. In: Labate BC, Cavnar C, editors. Ayahuasca Shamanism in the Amazon and beyond. New York: Oxford University Press; 2014. pp. 159-181. DOI: 10.1093/acprof:oso/9780199341191.003.0008

Chapter 4

People of Recent Contact in the Ecuadorian Amazon

Patricio Trujillo-Montalvo

Abstract

The Waorani are an ethnic group of recent contact who preserve traditional cultural practices of people who inhabit the tropical jungles and are characterized by being clanic families of hunters and gatherers, highly mobile and nomadic warrior groups that inhabit a wide expanse of Ecuadorian Amazon forests (about one million hectares) located in the Yasuní National Park (YNP), a unique humid forest ecosystem in the world, which they consider their territory. Their language is Wao Tiriro, the word wao meaning "human" and Waorani "human beings." It is their form of self-identification with their ethnic neighbors, whom they call cuwuri or strangers.

Keywords: Waorani, Amazon, civilization, modernity, contact, ethnography

1. Introduction

The images created around the Ecuadorian Amazon jungles and their inhabitants are the product of the confrontation between various logics of thought, different worldviews, ethos and symbols. This confrontation has been carried over to the present day, and in modern stages, it is much more valid since the appropriation of this space by different actors results in the systematic consolidation of different power groups that, when confronted with each other, generate conflicts of different kinds and violence in the region, becoming one of the references for the creation of the local identities.

If the confrontation of these logics results in the appearance of new actors, we have the Ecuadorian Amazon converted into a true laboratory of ethnic constructions and disintegrations. An ecology, where various groups, representing national and colonial states and even tribal movements, fight to consolidate this region as their space of power. The clash of these two logics: an extractive market economy (gold, spices, rubber, oil and coca) and the other of resources for survival (hunting, fishing, gathering and planting), have generated that the Amazon jungle is seen as a wide region of inexhaustible resources, of invisible inhabitants to be conquered, Christianized, civilized and transferred to modernity.

The passage from the wild subject to the modern subject is seen so clearly drawn in these areas, where the different regions and their inhabitants go from representing spaces of cultural reproduction to wasteland regions, without owners, without God, law or order. If 40 years ago, the ideal of development on a global level was to create a humid tropical forest, the source of food for the planet, and to civilize its inhabitants and incorporate them into the modern world. In the last decade, this ideal has been linked to maintaining tropical humid forests in a natural state, conserving them as the lungs of the world and their inhabitants as the last modern savages on earth, as the guardians of the tropical humid forest. Contradictory paradigms of the new modernity that use the inhabitants of the tropical forest to justify the continuous errors of development projects, executed from the civilized north to the underdeveloped periphery.

2. Waorani: savage and civilized

The term Wao (wao tededo) means human and Waorani means (true) human beings. This is how this cultural group identifies itself and creates its ethnic borders in relation to other groups, which it calls "cowode"/cowuri/or the non-Waorani, "the strangers, the outsiders, the others." In specialized literature, you will find different ways of writing this term, like this: cowode, cogouri, cuwuri, cowuri, cohuori, etc., there are many reasons for this, but perhaps the most real is that the first ethnolinguistic studies were carried out by North Americans and the other phoneticizations depend on the mother languages of those who have worked with the Waorani. The literal translation of the term is strange, but it has other "charges" in the message such as savage, cannibal and murderer [1–3].

The Waorani language (Wao Tededo, Tiriro or Terero) has no affinity with any other language or parent linguistic group in the Ecuadorian Amazon and is considered an isolated language. In fact, the toponyms of this space, known since the time of the missions, do not correspond at all to the Wao Tededo language. According to the story of several old warriors (Pikeneni), there were different dialects among the different clans, the same ones that are divided between two large groups IROMENANI or those upstream and ENOMENANI or those downstream of the Napo River [4]. An interesting argument to relate this status of supposed marginality, in the ethnic context in which they developed, is linked to the fact that until the mid-twentieth century, the area was dominated by groups or tribes of Tukano linguistic affiliation, especially located toward the left bank of the Napo and by those of Sapara and Kichwa affiliation, starting from the right bank. The variety of names (anthroponyms-toponyms) of the different rivers and sites that have been preserved in early references by missionaries and travelers prove this hypothesis: Napo (Doroboro), Tiputini (Guiyero), Yasuní (Dicaro), Tivacuno (Peeneno), Curaray (Ewengono), Tiguino (Bataboro), Cuchiyacu (Menkaro), Cononaco (Baameno), Nashiño (Gabaro), etc., against which the Waorani have their own linguistic name.

Later oral and written references allow us to assume that, at least for a certain time, the surroundings of the mouth of the Tiputini were inhabited by some Omagua or Sapara tribes. At the time of the "rubber patterns" (1910–1930) apparently the populations located in the same environment, between the Tzapino, Villano and Curaray rivers, were not Waorani but Sapara [5]. Apparently, the Waorani warrior clans eliminated the Sapara families, to later occupy these sites and incorporate them into their territory. In fact, a route allowed rubber seekers to cross from Curaray to Napo, passing through Cononaco, Nashiño, Yasuní and Tiputini, which at this time appeared populated by clans of Sapara descent.

The rereading of the innumerable ethnographic data (writings of travelers, missionaries) before contact with the West indicates that this group was classified as extremely violent, highly mobile, fighting with any stranger who had entered their territories. This created the imagination of savagery, especially among the Kichwas, who called them "Aucas" (savage or uncivilized in the Kichwa language). "Auca" was the name that identified them for a long time, until the Summer Linguistic Institute (SLI) reinvented for the West as Waorani [6].

The Waorani are a group that, from the tradition of Amazonian studies, is considered a tropical jungle culture [1, 7], mainly because it maintains the following basic characteristics:

a. Generalized war fulfilled the function of social organizer; it was the most important socio-cultural and control institution.

b. Hunter-gatherers with high mobility and little horticulture. The use of forest resources depended on the seasons of fruit and animal reproduction in the forest.

c. Clan groups or extended families made up of a complicated kinship designation. Each group was dispersed and maintained little contact related to the lack of centralized power. The chief was named or constituted in a circumstantial way, the power was egalitarian and related to the levels of symbolic power and the best warrior was the clan chief.

d. Linguistic groups, oral cultures that recreated their entire worldview based on myths and legends that are transmitted from generation to generation.

e. Territory conceived as a large space of resources. Their houses were built in hilly areas of mainland forests, allowing them to see their enemies, attack and flee, away from the main rivers, avoiding any contact with strangers.

2.1 Contact and modernity

Before formal contact with the West (1956), in the oral memory or Wao tradition, two eras or times in which the different groups lived are remembered: (1) time of tranquility or peace (waemo eñere), and (2) times of war, escape or group diaspora (piinte eñere). These stages shaped the social history of the Wao group, its development and, above all, its survival strategies against the "other groups" called "cowuri" and other enemy family groups called "warani or non-nanikaiboiri." In times of peace, they remember stability and stages of sedentary lifestyle, where the development of clan groups (nanikabos) was based on horticulture and hunting, family settlements remained from 1 to 3 years in the same place that was considered as its area of action or territory. On the contrary, the war stage was characterized by high mobility, search for safe places and family survival based on hunting and gathering seasonal fruits of the forest; horticulture was secondary and settlement sites did not last more than 2 years. Months due to the fight from the insistent intragroup war [1].

Nanikabo is the social and economic unit that brings together several extended families (Guiquitaris, Wepeiris, Piyemoiris, Nihuairis, Kempeiris, etc.), linked by kinship, whether blood or fictitious (alliances). Parental alliances were fundamental for the development of the group, since survival against their enemies depended on them. The Waorani maintain a complicated kinship relationship since they are polygamous groups in both ways: polyandry, woman with several men or polygyny, man with several women. Another cultural institution was war, the fundamental basis of social relations between the different Waorani clan groups. The same one that would become the main socio-cultural institution because, around this exercise of power, complex strategies of alliances and mythical constructions that fed the oral culture of the Waorani were developed and consolidated. In fact, during the war period, "death with spears" (tapaca wente) was the main social mechanism that regulated demography and symbolic relations between clan groups [4].

The first contacts between Western society and the different Waorani clans were violent and are described in several accounts of travelers who entered this territory and compiled several stories about the bloody war they waged against the rubber tappers, who, between the decade of 1910 and 1930 of the twentieth century they crossed the Curaray, Villano, Tiputuni, Shiripuno, Cononaco, Yasuní and Napo rivers, in search of raw materials and enslaved people for the haciendas of Peru and Brazil. The "hunt" to which the lower Cononaco-Baameno and Curaray groups were subjected, especially Baiwairis (people of Baiwa), is present in the memory of many warriors today, memories that trace back the cruelty with which many Waorani were enslaved and sold in distant haciendas., the majority never returned, so the image of the "cowuri" as a murderer who "eats or takes away" people begins to be conceptualized among their Wao culture since that time.

Subsequently, several confrontations are described when the oil company Dutch Shell Oil Company (1940) enters Waorani territory, especially the Arajuno River area, and establishes a camp to carry out seismic prospecting operations throughout the area. Moipa, Iteka and Guikita, are the warriors who defended this part of the territory, incessantly attacking the oil workers, resulting in several deaths between the two sides. Blonberg, in his book *The Naked Auca* [5], accurately recounts all these events that contributed to the creation of the image of the "ferocious Aucas," groups reluctant to contact and civilization. As we pointed out, Aucas (naked savage in Kichwa) was the word that was used for a long time to designate or identify the Waorani clans that attacked and defended their territory until 1956, a date that marked a new relationship with the West, since specifies the contact mediated by the Summer Linguistic Institute-Wycliffe Books Translators SLI-WBT (Summer Linguistic Institute), an evangelical organization in the United States, the same one that had been contracted by the Ecuadorian State to "civilize," "pacify" and "evangelize" the Aucas, who had caused so much trouble in the area of the Ecuadorian Amazon provinces of Pastaza and Napo.

2.2 The Christian God

The SLI entered the Ecuadorian jungles in 1953, forming its first operation center in Shell, near Puyo. Until then and for many years, as has been explained, there had not been a peaceful relationship between the Waorani and the "cowuri." For months, the missionaries prepared the meeting, dropping from the air (using devices from the Friendship Air Mission-Aircraft Fellowship Mission) steel materials (axes, machetes, knives, pots, etc.), intended as gifts for this operation. It was called "Operation Auca." Contact was made after a small missionary plane landed on the "beaches" of the Curaray River and five of them (Nate Saint, Jim Elliot, Pete Fleming, Ed McCully and Roger Youderian) managed to establish the first camp, which they called "palm beach" or "playa de palma." At this site, they wait for the arrival of the group of Auca, to whom they had dropped gifts from the air. After a few days, a small group appeared, with which they tried to make peaceful contact [8].

On January 8, 1956, the missionaries were speared to death. A fact recorded in documentaries, magazines and books, since when Ecuadorian army forces entered to rescue the bodies, they found film and photographs that the missionaries had taken of their alleged murderers. Later, Minkaye, one of the actors in the event, reported that there was a confrontation between the two groups, and that the missionaries had shot one of the Waorani, starting the confrontation and the subsequent massacre with spears [1].

In 1958, Raquel Saint (SLI) and Ellizabeth Elliot (Christian Mission in Many Lands), sister and wife, respectively, of two of those murdered, returned to the United States in order to raise money to continue with "Operation Auca." Indeed, they did so, and they returned to Ecuador with much more financial support to finish their project. Later, Raquel Saint met a Waorani woman, on one of Carlos Sevilla's properties (Hacienda Ila), called Dayuma, a young woman who had fled with other members of her family, when her father Kaento was killed by Moipa [3, 8].

The SLI quickly copied it and converted it. Raquel Saint takes Dayuma to the United States and there presents her as the first Christianized wao. Saint quickly learned the Waorani language (wao tededo) and together with Elizabeth Elliot returned to the Tiweno river, with the mission of looking for Dayuma's family, telling them about "the goodness of the new God, telling them not to fight anymore, it was time to fight to change life he did so," and his return meant the main anchorage for his group to move to the new area that the SLI missionaries had prepared, the protectorate. Dayuma became the emissary of the SLI and the evangelical God, trying to get all her relatives to gather in a single area and be pacified. From that moment on, the SLI becomes the protector of the Waorani and maintains intense control and zeal over their lives [3, 8].

Later, with other Waorani such as Zoila (Wiñame), Minkaye, Kimo and Dawa as interlocutors, a select group of anthropologists and linguists (James Yost, Catalina Peeke, Patricia Kelly, Raquel Saint) from the SLI mission entered, to learn the Waorani culture and language, translate the Bible into their language, evangelize and civilize them. From this moment begins the modern history of the Waorani; they would never again be like their warrior grandparents from the tropical jungle; it is the definitive step to a new world. However, it was from 1968 onwards that the true process of reduction of the Waorani by the SLI began. Almost 80% of the clan groups are convinced to abandon their traditional lands and move toward the so-called Tiweno, near the famous Palm Beach or Playa de Palma, currently Toñampare [1, 3, 4, 8].

Finally, in 1969, the Ecuadorian State handed over 1600 km^2 to the SLI, known as "the protectorate," which is where they concentrated and began the pacification of the majority of the Waorani clans. By 1972, most of the clans had abandoned their traditional territory of residence, moving toward the so-called "protectorate," which was located at the southwestern end of their territory. Many were transported by plane with their belongings, with the move of clan chiefs, thousands of kilometers were left empty and uninhabited [3, 8].

However, not everyone migrated to "the protectorate," some groups stayed in the jungle, continuing with their traditional way of life. The groups of the so-called "lower Cononaco" that lived near the Yasuní and Nashiño rivers: Nampaweiris, Wepeiris, Waneiris, Baiwairis, Kempereiris, never moved to the SLI village, remaining in their territories for several more years without contact. This is how the historic settlement pattern of the Waorani (transhumance) was modified with the "evangelization" and the simultaneous process of "civilization" that they were subjected to by the "missionaries" of the SLI. Currently, the majority of Waorani population groups settle in the area known as the protectorate [4].

2.3 The last warriors of the Ecuadorian Amazon

Tageiri-Taromenane are the last groups of men and women in the Ecuadorian Amazon (and possibly the global Amazon) who do not remain in close contact with national society, either by their own decision in accordance with their cultural norms, or by flight from their enemies and the effects that Western civilization has on their lives and cultures. They have Wao Tiriro linguistic belonging and are characterized by high mobility and uxorilocal parenthood, a form of matrilocality that evidences the return of Waorani-Tagueiri women to the territory where their parents were born [9]. This cultural trait makes them one of the last cultural groups of tropical humid forests with traditional characteristics in Ecuador. However, relations with their cultural and linguistic relatives, the Waorani, go through complex dimensions of conflict over territory. To this relationship must be added symbolic components related to a form of intragroup violence and war in which a series of episodes of deaths and revenge have configured complex war geography between Waorani and Pueblos Indígenas en Aislamiento Voluntario (PIAV) [4].

The Tagueiri were members of the Waorani affiliation who separated more than 40 years ago and moved among a large jungle area that includes the Nashiño (Gabaro), Tivacuno (Peeneno), Shiripuno (Keweriono), Tiwino (Menkaro), Cononaco (Baameno) rivers) and Curaray, which forms a corridor of high seasonal mobility. It seems, according to testimonies collected in this investigation, that they were absorbed by another PIAV group known as "taromenane" [1, 4, 9].

These groups are described as hidden or uncontacted peoples who live in the areas of the Waorani territory and the Yasuní National Park and are defined as extended Waorani clans or families who have decided to live away voluntarily due to pressure from their ethnic enemies or due to the pressure that the expansion of the agricultural and oil extraction frontier exerts on their mobility territories: "They are groups that maintain high spatial mobility, which allows them to survive with a certain independence and isolation, compared to the way in which Waorani families lived before contact with the SIL, called by informants as 'durani bai'" [4].

This article proposes that the Tagueiri-Taromenane clan families could not be classified as hidden peoples, much less "without contact," since there is evidence of kinship relations, exchange and symbolic alliances with other Waorani clan groups, such as the Baiwairis. (the Baiwa people), Kempereiris (the Kemperi people) and Kairis (the Kay people), three of the Waorani families that control and inhabit part of the territories of the Yasuní National Park and, specifically, the areas of the Cononaco rivers (Baameno), Nashiño (Gabaro) and Yasuní (Kawymeno), where there is evidence of the presence of the PIAV [4].

The spaces of contact, peaceful coexistence or not, between PIAV and Waorani families, are interpreted on several levels. At the first level, it is related to (a) exchange of forest products, for example, hunting prey for cassava, banana for handicrafts or for city products such as machetes, axes, salt, pots, etc., (b) with phenomena symbolic, referring to exchanges at festivals for good hunting, potential alliances and marriages and (c) with the war with Niwairi and Babeiri enemies, especially those who maintain a warlike relationship with these groups and who have maintained a violent relationship over several years., with stages characterized by times of peace and times of war and revenge [4].

The second level of contact would be related to spaces of shared territoriality with Waorani families. This is the most conflictive space because it produces competition for resources that are scarce due to demographic pressure and also for territory. The Tagueiri-Taromenane are characterized by high mobility and the use of large forest spaces in order to obtain resources for their survival. They imagine that large space of jungle as their territory, a space that in geographical coordinates would limit between the banks of the Curaray and Napo rivers.

The situation of the Tagueiri-Taromenane is complex because they are pressured or surrounded on several fronts. The first refers to the expansion of the agricultural frontier and the shared and conflictive use of the same hunting and gathering territory by Waorani clans. The second corresponds to the oil extraction activities that have expanded in the area of influence them and that have facilitated the construction of roads that at the same time allow permanent settlements of settler-peasant groups near YNP (Los Reyes, Armadillo, Hormigueros), contacted Waorani families (Miwaguno and Yawenpae) and Shuar-settler families (Tiwano), who increasingly put pressure on the territories through which these families move (especially those located between the Tivacuno-Armadillo-Cononaco Chico-Menkaro area) [4].

The future of these families is complex, especially after the massacre in April 2013, their mobility spaces are increasingly reduced and the relationship of vulnerability and conflict, resulting from an accelerated reduction of their living spaces (territory) and potential time for an encounter with enemies is getting shorter and shorter. A new confrontation with warriors from enemy Waorani families could eliminate them completely: their extinction is a real possibility. Finally, this article considers the option of a slow but sustained fission between the Tagueiri-Taromenane families and the Waorani families, for example, Kairis, Kemperiris and Baiwairis clans. Libidinal exchange has been one of the survival strategies of Amazonian cultures; perhaps that is what is happening in the vast jungle areas of Kewerioro, Baameno, Nashiño, Kawymeno and Curaray in the Yasuní National Park and the Waorani land.

2.4 Cultural change?

The control of demographics in the Waorani culture was closely related to war and the extensive deaths between clans that it caused. "Murder or death with spears" was one of the main drivers of population growth; this could be measured according to data that James Yost [10] has before contact:

- 44% of Waorani died from intra-clan murders by other Warani or enemy but families
- 12% illness or witchcraft
- 10% were murdered by cowuri (rubber farmers, oil workers and Kichwa indigenous)
- 12% were kidnapped by landowners
- 8% fled or hid in various places
- 6% snake bite

- 5% accident
- 2% old age
- 1% unknown

At the time of pacification and subsequent sedentaryization in the protectorate, the new way of life brought with it an accelerated process of population increase between 1958 and 1990. According to data from various sources [1, 10]:

- 1958 there were approximately 500 Waorani
- 1980 there were 658
- 1982 there were 715
- 1990 there were 1157
- 1993 there were 1282
- 1996 there were 1580
- 2001 there were 1898
- 2010 2000
- 2020 3050

Consequently, the growth rate in a period of just over 50 years was more than 300%. The population increase is related to the rapid social transformation of the Waorani; basically linked to [1, 4]:

- The suspension of intra- and extra-group violence and cycles of inter-clan revenge.
- The change in settlement patterns, greater sedentarization, little spatial mobility, construction of towns.
- Reduction in the practices of geronticide (death to the elderly) and infanticide (death to children).
- Abandonment of polygamy and traditional marriage alliances (Zororate-Levirate) due to the emergence of modern practices (Cowuri-style marriages).

Starting in the 1970s of the twentieth century, a rapid process of reconfiguration of the spaces of this group began. From nomadic groups, they become groups with permanent and semi-permanent settlements, which definitively and forever change their previous relationship with the forest. Currently, the Waorani settlements are located in three areas: the westernmost region that corresponds to the so-called “protectorate,” the same one that was defined by the grouping of families initiated by

SLI. To the northeast of the ancestral territory on the banks of the Yasuní rivers and their tributaries, there are also small groups and, finally, the areas of roads penetrated by the oil industry or block 16 and block 21.

Now, before 50 years of contact, the Waorani territory and the Yasuní National Park (YNP) are divided into four spaces:

1. A sector dominated by the Kichwa communities located between the banks of the Napo River and the Tiputini River in the north, and the Curaray River in the south.

2. The colonized areas on the axis of the oil route are known as "Auca road."

3. The Waorani territory in block 16.

4. The entrance road to block 16, shared by Kichwa and Waorani.

3. Conclusions

The contact with national society has been characterized as aggressive, since "homicide and raids mark the uncrossable border between the cowuri and Waorani worlds." The social solidarity and cultural unity of all the Waorani are marked by the true "human beings," compared to the cowuri, the "non-human-enemy-cannibals," however, this interior-exterior dichotomy is also found within the Waorani world. The introduction of the Waorani into the market economy has intensified in recent years. Young people work in oil companies, and with the money, they buy objects of Western culture. This tendency, according to Rival, is the result more than the dynamics of capitalist commodification, of the influence of the formal institution such as the school "that represents modernity and its material manifestation, "Western products" [1].

The relationship with national society has been traumatic for the Waorani people and always violent. There has been a silent war that has caused several clashes and deaths on both sides. The conflicts with settler groups, other indigenous groups like Kichwa and Shuar, oil companies and missionaries have generated an imaginary of violence and savagery among the Wao people... "these are the last savages... they are brave Indians, you have to go there carefully, they are murderers of birth, few civilized and dangerous" [4].

This imagination has created, in contrast, excessive zeal and care on the part of foreign and national conservationist environmental groups, which is why the defense of the Waorani has become the slogan of the struggle of several NGOs, who see in this group the hope for the conservation of large spaces of tropical humid forest.

For the Waorani of recent contact, the process of relationship with the cuwuri reproduces a very characteristic relationship with their jungle environment, exercising the practice of the warrior or collector. Waorani praxis identifies the actors with whom it is important to maintain a relationship, as a strategy in that conception of territory of abundance and limitation is where external actors are incorporated, and with whom the same relationship is established as with the jungle, whether oil workers, settlers, missionaries, formal and informal tourism operators, who are asked for their "contribution" with products or services [1, 4, 9].

In the historical context, the attacks generated during the first years of the previous century occurred against indigenous laborers of haciendas that are located on the banks of several rivers, such as the Napo in its upper and middle course, Curaray, and others. Next, and in the context of oil exploration, the attacks occurred against oil workers, whose victims belonged to several companies that carried out their activities in Yasuní.

In this context, the State's limited understanding of diversity has led to the assumption that all indigenous peoples have a community character, this being inapplicable in tropical forest societies. The relationship of collecting the resources provided by the actors that operate in their territory, being no longer a "relationship of assistance" but rather a relationship in which "collection" practices or actions are identified and where the Waorani actors establish themselves as "great men" in this new gathering jungle, different from the traditional space of the elderly where the defense of the territory, and the experience of the warrior code established the leaders and socially recognized characters [9].

These great men are Waorani who were children at the time of contact or were already born in the post-contact era, while the elderly, who lived in a traditional way, still maintain memories around the "duranibai" where the warrior ethos prevailed, which configured them as providers and guarantors of family survival in the territory, which today is reduced to a monetized economy dependent on a salaried job, the dramatic change from being Waorani to being an impoverished citizen [1, 9].

Acknowledgements

Special thanks to the Pontifical Catholic University of Ecuador for funding the article and to the Waorani friends who for more than 20 years have welcomed me into their homes as part of their families.

Conflict of interest

The authors declare no conflict of interest.

Appendices and nomenclature

SLI	Summer Linguistic Institute
YNP	Yasuní National Park
NAWE	Nacionalidad Waorani del Ecuador-Waorani Nationality
PIAV	Pueblos Indígenas en Aislamiento Voluntario-Indigenous People in Voluntary Isolation
YNP	Yasuní National Park, Waorani land and principal rivers (home range): Napo, Tiputini, Curaray, Yasuní, Cononaco, Nashiño, Shiripuno, Tivacuno, Tiguino

DOI: http://dx.doi.org/10.5772/intechopen.1003633

Author details

Patricio Trujillo-Montalvo
Pontificia Universidad Católica del Ecuador, Ecuador

*Address all correspondence to: pstrujillo@puce.edu.ec

References

[1] Trujillo P. Bito cowuri, Boto Waorani, la fascinante historia de los Waorani. Quito: Fundación de Investigaciones Andino Amazónicas (FIAAM); 2011

[2] Peeke C. Gramática Huaorani. Cuadernos Etnolinguisticos. Quito: Instituto Linguistico de Verano; 1979

[3] Wallis E. The Dayuma Story. New York: Harper and Row; 1975

[4] Trujillo P. Salvajes, civilizados y civilizadores; la Amazonía ecuatoriana, el espacio de las ilusiones. Quito: Fundación de Investigaciones Andino Amazónicas (FIAAM)/Abya-Yala; 2001

[5] Blomberg R. Los aucas desnudos: Una reseña de los indios del Ecuador. Quito: Abya Yala; 1998

[6] Boster J, Yost J, Peeke C. Rage, revenge and religion: Honest signaling of aggression and nonaggression in Waorani coalitional violence. Ethos;**31**:471-494. DOI: 10.1525/eth.2003.31.4.471

[7] Clastres P. La cuestión del poder en las sociedades primitivas. In Investigaciones en antropología política, editado por Pierre Clastres, 109-116. Barcelona, 2003. 1981

[8] Elliot E. The Savage my Kinsman. New York: Harper and Row; 1977

[9] Narváez R. Intercambio, guerra y venganza: el lanzamiento de Ompore Omehuai y su esposa Buganei Caiga. Revista Antropología: Cuadernos de Investigación. 2017;**16**:99-114

[10] James, Yost J. Twenty years of contact: The mechanisms of change in Huao (Auca) culture. In: Whitten NE Jr, editor. Cultural Transformations and Ethnicity in Modern Ecuador. Illinois: University of Illinois Press; 1981. pp. 677-704

Section 2

Modern Life

Chapter 5

Difficulties in Accessing Medical Surgical Care in the Amazonas

José Emerson Souza, Cleinaldo de Almeida Costa, Gabriel Peixoto França and Nivaldo Alonso

Abstract

Despite over half of Amazonas's inhabitants residing in rural areas, few specialized health services are available to them. Most advanced health services, including hemodialysis, are concentrated in the state capital, Manaus, leaving those in rural areas without access to proper healthcare. The population of the interior is devoid of units of health. This is partially due to the region's challenging geography, making it difficult for rural populations to reach Manaus. All surgical hospitals in cities surrounding the capital are publicly funded, with no private healthcare options available in rural Amazonas. The Lancet Commission on Global Surgery, an international initiative, has proposed a model for analyzing the health system, especially the surgical system, to achieve universal access to safe surgery and anesthesia by 2030. Funding is a crucial factor in making this possible and providing better access to healthcare for all. A comprehensive analysis of the health system in the state is necessary to guide public policies, optimize future healthcare investments, and improve access to clinical and surgical treatments for the population.

Keywords: health systems, health care infrastructure, health care quality, health provision, health impact assessment, essential health indicators evaluation

1. Introduction

In the late 1980s, the "Health Reform Movement" was launched in Brazil to oppose the military dictatorship. As part of this movement, the single health system, known as SUS, contemplated by civil society, is today a successful reform model and an example of a health system for Latin America [1]. It played a crucial role in the re-democratization of Brazil and the restoration of citizens' rights [2]. Reforms in health system governance and the growth of primary healthcare (PHC) have led to considerable enhancements in health service coverage, accessibility, and outcomes [3–5].

The Brazilian health system has a mixed coverage, mixing elements from the public and private spheres, not only in terms of service provision but also in terms of funding [6]. It comprises a diverse range of primary, secondary, and tertiary health services. These services are predominantly funded by public financing, with private involvement mainly in the secondary and tertiary sectors. Either companies or users themselves pay for private health plans.

The organization and delivery of health services are divided into Primary care, Secondary care, and Tertiary (hospital) care. The first is the public sector, where the government provides financing and services at municipal, state, and federal levels, including healthcare for military employees. The second sector is the private sector, which includes for-profit and non-profit organizations. These services can be paid for with public funding or directly by the user. The last sector is private health plans, which offer various premiums and tax subsidies. Although distinct, the public and private sectors are interconnected since users can access both, and financing can come from either the government or the user. The private health sector provides additional hospital and outpatient services alongside the public system. Most of the funding comes from public sources, with other private contributions. Health plans are mainly used by employees of both public and private companies [2].

The primary health care model intends to offer universal access, and the Family Health Program (Programa Saúde da Família -PSF) has been the primary strategy for structuring municipal health systems. It consists of a program composed of several primary health care units, containing medical and nursing professionals, distributed among the municipalities, aimed at primary disease care, such as systemic arterial hypertension, diabetes mellitus, and care for pregnant women, among others. The PSF's primary goal is to organize small health units in each city's health districts that cater to families and communities. These units focus on preventive measures and health promotion. Additionally, they act as a triage system for secondary and tertiary health levels.

One of the significant challenges in the healthcare system is at the secondary level. This level comprises units responsible for conducting various types of medium and high-complexity exams. These services have a limited supply and are mainly for the private health plan sector [7, 8]. Access to equipment in the public healthcare system is limited, with only 24.1% of computed tomography devices and 13.4% of magnetic resonance devices available for public use. The situation is even more challenging at the tertiary level, where highly complex and costly surgical procedures are performed. Public university hospitals and private hospitals contracted by the government are responsible for providing these services, adding to the challenges [9].

The State of Amazonas is one of the seven states in the North region of Brazil. It is located in the central part of the country's northern region and is part of the Amazon Biome. It is part of the Legal Amazon, together with the states of Amapá, Acre, Tocantins, Rondônia, Roraima, Pará, north of Mato Grosso, and west of Maranhão. It borders Venezuela and Roraima to the north, Colombia to the northwest, Pará to the east, Mato Grosso to the southwest, Rondônia to the south, and Acre and Peru to the southwest (**Figure 1**). It is the country's largest state in terms of land area, covering 1,559,161.682 square kilometers and 62 municipalities. It also has 97% of its forest cover preserved, in addition to about 12% of the total fresh water in the world, being considered therefore the largest freshwater reservoir on the planet [10]. The total population, according to the last census carried out in 2022, is 3,952,262 inhabitants, representing about 2% of the country's population, who live primarily in urban areas, with more than half (2,054,731 inhabitants) residing in the capital and has the second-lowest demographic density (2, inhabitants per km^2) among the Federation Units [11]. This entire population is distributed in 9 health regions, described in **Figure 2**.

The vast expanse of Amazonas state poses significant challenges for its healthcare system. Transportation between cities is primarily through rivers, which can take hours or even days. Small aircraft are also used but are expensive and not accessible to most people. Only a few cities have highways connecting them to the capital [12]. Because of the diverse population distribution in the state, health management

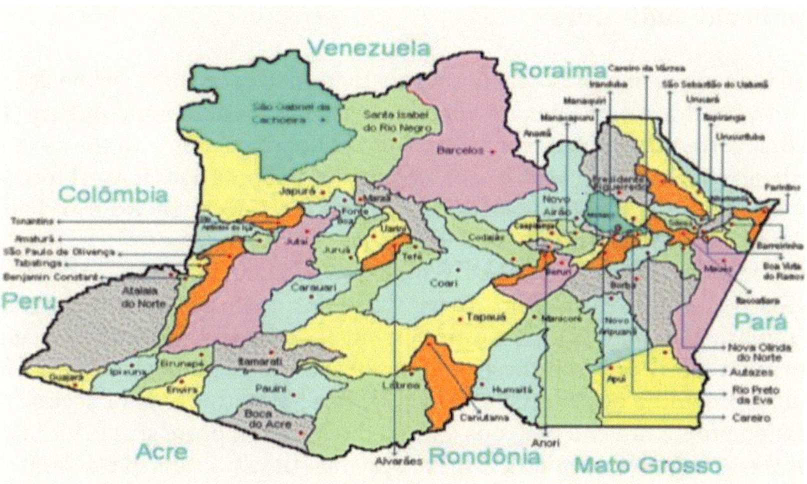

Figure 1.
Map of Amazonas with limits.

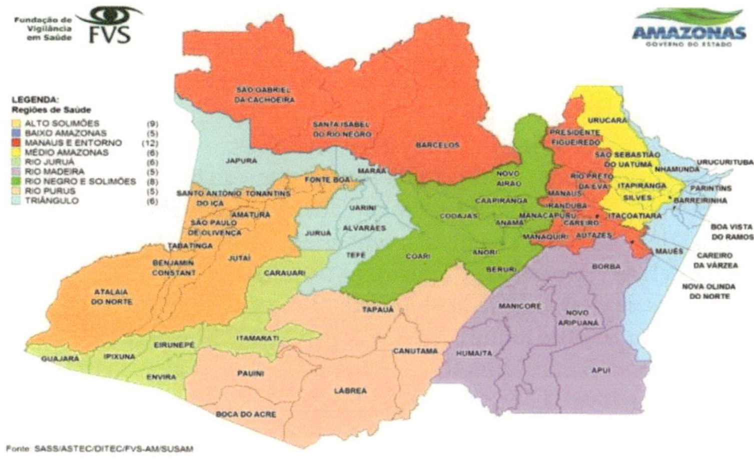

Figure 2.
Map of Amazonas distributed by health regions and their municipalities.

requires a resourceful and expensive approach. This involves transporting patients by boats and planes, utilizing telemedicine, and relying on surgical missions.

Although there are hospitals in rural communities throughout the state, there is a need for more advanced facilities for major surgeries and specialized exams, like magnetic resonance imaging [10]. Complex hospital units are limited in their ability to provide essential surgical procedures and emergency surgeries due to a shortage in the surgical workforce. Surgeons, obstetricians, anesthetists, and surgical nurses are particularly affected. Despite their skill, they often lack the necessary equipment to perform complex surgeries. This is due to a lack of specialized professionals, infrastructure, and access to material and financial resources. As a result, service provision is often precarious [13].

For many, access to secondary and tertiary-level hospitals requires days of travel and high transportation costs, which is often unfeasible for a large part of the population.

2. Major health indicators

Brazil is an upper-middle-income country that, despite its universal health coverage scheme, has dramatic regional variations in healthcare access and quality [14, 15]. Within Brazil, the state of Amazonas is one of the poorest states with the most significant challenges in accessing health care, mainly due to limitations in workforce and infrastructure [16]. The state's inadequate diagnostic services may worsen challenges in providing safe, timely, and affordable care.

The Lancet Commission on Global Surgery (LCGS), an international initiative, has proposed a model for analyzing surgical systems. In its Global Surgery 2030 report, published in May 2015, the Commission describes its vision of universal access to safe surgery and anesthesia and establishes that funding is essential for its viability, providing better access to all [17]. To achieve this level of access, the Commission proposes a two-step analysis method: first, it proposes six indicators to assess the strength of a region's surgical system—timely access to essential surgical care, number of surgical teams, volume surgery, perioperative mortality rate, protection against impoverishment, and protection against excessive expenses [18]. These indicators aim to quantify the preparation and capacity of the Surgical System and the system's ability to protect patients against financial risks. Second, the commission proposes structuring the surgical system that considers regional characteristics and contexts in a particular way. The main components of this plan are: 1—infrastructure, 2—workforce, 3—provision of services, 4—information management, and 5—financing [18].

To assess the quality of a healthcare system, a qualitative and quantitative mixed methods Hospital Assessment Tool (HAT) was developed [19]. The HAT tool was developed jointly with the Global Surgery and Social Change Program and the World Health Organization (WHO). The instrument assesses the infrastructure of a surgical system, service delivery, workforce, information management, and funding through hospital interviews, reviews of medical records and surgery books, as well as interviews with unit directors and some providers responsible for surgical care (surgeon, anesthetist, obstetrician, nurse, and hospital managers). An initial HAT pilot project was conducted in Cape Verde, Ethiopia, and India. The tool was then adjusted and validated by 18 experts (Delphi consensus) [20].

The tool has been recently used in Brazil's largest state, Amazonas, which also has more difficult logistical characteristics, to identify priority areas for system improvement and health policy changes, as perceived by patients and local providers. The tool's implementation involved a partnership between the University of the State of Amazonas (UEA), Harvard Medical School, and the University of São Paulo (USP). Understanding this region's deficiencies and strengths helped identify the main gaps in the delivery of medical services in the state.

2.1 Health infrastructure

A peculiar characteristic of the state of Amazonas refers to its geography since it is crossed by several rivers. This makes access to cities very difficult. Because of this, the general infrastructure of hospital units in all cities is quite deficient. We can see a direct reflection of this geography when we observe that all hospital units in these cities (except for the capital) are public, with no private medical assistance at the hospital level outside the capital Manaus.

Figure 3.
Basic health unit working as a hospital.

These hospitals generally provide clinical, surgical, obstetric, and pediatric care. In some cases, such as in the city of Careiro da Várzea, located close to the capital and with an estimated population of 30,846 people, only primary health care is provided, with no capacity to perform surgical procedures (**Figure 3**).

Our study shows that district hospitals in low- and middle-income countries often experience water, oxygen, and electricity shortages, whereas our findings indicate better infrastructure [21–23]. According to the survey conducted, almost 95% of the hospitals located in the Amazonas interior are equipped with piped water and a consistent supply of electricity. However, only one hospital provides channeled oxygen, while others rely on cylinders. Essential radiology services are available 24/7 in 83.3% of the hospitals, while half of the units always have ultrasound services available [13]. Computed tomography or magnetic resonance imaging equipment is available in only one hospital [24].

The biggest problem found regarding infrastructure refers to the number of hospital beds. Among the 18 hospitals surveyed, there were a total of 620 beds available, with an average of 34 beds per hospital (ranging from 4 to 102). None of the hospitals included in the study had an intensive care unit (ICU). Most hospitals had 1–2 operating rooms (as shown in **Figure 4**). Three hospitals (16.7%) had no operating rooms [13].

2.2 Medical workforce

From 1970 to 2011, the population of physicians grew by 530% in Brazil. In that same period, the Brazilian population grew by 104%. The speed of evolution of the doctor/inhabitant ratio will be further accelerated with current government policies that have considerably increased the number of medical vacancies [25].

In 2011, a resident of the capital of any state in the South and Southeast had four times more doctors than a resident of any other region in Brazil. The federal government publicizes that the problem is the general lack of doctors in Brazil, expressed in

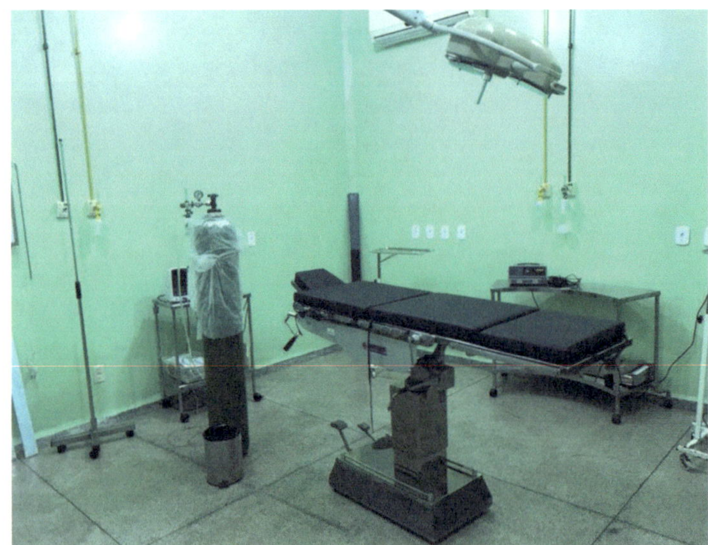

Figure 4.
Typical remote hospital operating room.

a single rate for the entire country. But the problem is another: the inequality in the distribution of physicians, with a super concentration of physicians in the private sector and different cities and regions [26].

The differences stand out even more when the capitals are separated from the cities in the interior and when the population strata group municipalities. For example, 65.9% of Brazil's physicians practice in capitals, while only 24% of the national population lives there [27]. An extreme example is that of Amazonas, where 93.1% of doctors are found in the capital, Manaus, which, in turn, is home to just over half of the approximately 4 million inhabitants of the state. Likewise, only 56% of practicing physicians in the North completed residency training, compared to 78% of physicians working in the country [27]. Of the 4844 doctors in Amazonas, 4508 are in the capital, and 336 (6.9%) serve 62 municipalities spread over an area of 1.57 million km^2. Doctors from Amazonas represent 1.1% of the total number of doctors in the country [28].

In Brazil, the government's health scheme (SUS) is one of the world's largest public health care systems, aiming to provide universal health coverage to all Brazilians [27]. However, the major problem is related to the workforce at regional levels. Availability of specialist care varies by region, as only 56% of physicians in the North complete specialist training compared to 78% in the Southeast [15, 27].

In Amazonas, when analyzing the data collected directly in the hospitals of each city, it is notable that the workforce, especially in the surgical specialties (surgeon, anesthesiologist, and obstetrician), is the biggest problem. The surgical workforce across the state of Amazonas remains limited, with a density in the sampled regions of 6.4 per 100,000 compared to a target of 20 per 100,000 [17]. This density is lower than Brazil's overall density of 34.7 per 100,000 and interestingly, it is much lower than the modeled estimate for the northern region of 18.42 and for the state of Amazonas itself which is 23.82 per 100,000 inhabitants, highlighting the need for validated assessments at the local level [15, 29]. Some of these differences could be explained by the fact that 90% of the surgical workforce is concentrated in Manaus, the state capital, while in the rest of the country, just over 50% of the surgical workforce is focused on the state capitals [29].

Surgical workforce shortages, particularly in resource-poor environments, often lead to healthcare providers performing procedures without adequate training. Due to a lack of anesthesiologists, some surgeons perform anesthesia for their operations. Another observation was that there are some situations where non-surgeons routinely practice surgery and non-physicians perform procedures, often needing more training or supervision.

2.3 Service delivery

It is possible to make important considerations regarding the available supply of health services in Brazil. The majority of hospital services are privately owned: 62% of establishments with hospitalization and 68% of beds in the country. An even greater concentration is observed in relation to units providing diagnostic and therapeutic support services—SADT (92%). In contrast, most outpatient units (78%) are state-owned [6].

National surgical volume is an important parameter to assess the quality of medical services provided and correlates with maternal mortality rates MMRs. The countries with a surgical volume above 5000/100,000 have the lowest MMR [18, 30]. Simply measuring the total volume of surgeries performed does not ensure that necessary surgeries are carried out safely and on time. However, it is generally assumed that countries with higher surgical volumes also have higher rates of emergency and essential procedures [31].

In Brazil, the public sector alone performs 4337 surgical procedures per 100,000 people, which is close to the suggested minimum of 5000 procedures/100,000 [15]. It is crucial to examine the quality and case mix of surgeries, as well as personnel appropriateness. In particular, selective overprovision of surgeries such as Cesarean section [32–34] may be harmful to patients and draw resources away from the provision of other necessary procedures.

The low concentration of specialist physicians is reflected in the low volume of consultations and surgical procedures performed in remote hospitals in the interior of the state. The average number of surgical procedures performed in the capital Manaus is about 60 times greater than in the rest of the municipalities, much lower than the target proposed by one of the LCoGS indicators, which is 5000 operations per 100,000 people [13, 18]. This demonstrates the almost total dependence of the state on the capital for surgical care.

2.4 Health information management

Since SUS was implemented, several tools have been built to meet the Public Health demand, constructed by collecting data from the various information systems. The main tool comes with the creation of DATASUS in 1991, whose data provides concrete justifications for a series of public policies in health. The information acquired from the different systems of information comes with creating a department where it is possible to store large amounts of data, which can support public health policies [35].

Health Information Systems (SIS) are instruments that, through the processing of data collected in health services and other places, support the production of information for a better understanding of problems and decision-making within the scope of policies and care in health [36]. In Brazil, when these systems have national coverage, they are called national data sources under the management of the Unified Health System (SUS), SIS with national coverage or national base, with custody of the data and software maintenance usually under the responsibility of the Ministry of Health [37].

An important point identified in the evaluation of data directly in the units was the observation of information that differed from that found in the database of the Ministry of Health (Datasus) [24], such as, for example, the description of surgical procedures performed in the hospital unit of the municipality of Careiro da Várzea, which functioned only as a basic health unit, not performing any surgical procedure. This highlights the importance of this database and the need for proper feeding of this database.

2.5 Financing

In Brazil, the financial transfer to health service providers is carried out by the level of government responsible for its management, which may be municipal, state, or federal. Assessing the quality of a healthcare system involves considering general healthcare spending as one of the main factors. The total health expenditures in Brazil are similar to those in other Latin American countries. However, public spending needs to be improved for such a broad health system, leading to an overwhelming burden on the users [1]. When we compare public health expenditures, Brazil has one of the lowest proportions, around 46%, while Latin America and the Caribbean average 51.28%. In upper- and middle-income countries, these expenses rise to 55.2%, and the Organization for Economic Cooperation and Development countries recommends that these expenses should be around 60%. It is important to note that although private health plans can compensate for low investment in public health, they can also be a significant financial burden for families, accounting for approximately 50% of private health expenses [38].

Regardless of which sphere performs the payment, SUS uses the same information system for Outpatient Services (SIA) and another for the Hospital Information System (SIH). Although the payment for the service provided is decentralized, in the specific case of hospital admissions, the processing of information by SUS is carried out centrally by DATASUS. Thus, the entire public system uses a single price list, defined by the Ministry of Health, for payment to health service providers [39].

In Amazonas, most remote hospitals receive funding from the government, varying according to hospital production and surgical procedures performed in the units. Therefore, there are no direct expenses for the user. However, costs such as commuting, abstaining from work, and purchasing medication cannot be measured using the tool (HAT). In most developing countries without a public health system, the expense of surgery acts as a roadblock for people looking to receive prompt surgical treatment. It is a significant factor in their decision not to seek medical attention [40–43].

3. Conclusions

To ensure that the population of Amazonas outside of Manaus has access to safe, timely, and affordable healthcare, it is essential to evaluate health indicators at the state level and develop a plan for implementing public health policies. This is especially important given the unique geosocial makeup of Amazonas and the limited state authority on health. It is also important to assess hospital infrastructure in remote areas to identify areas for improvement and understand the barriers the remote population faces in accessing surgical care.

DOI: http://dx.doi.org/10.5772/intechopen.1003082

Author details

José Emerson Souza[1*], Cleinaldo de Almeida Costa[1], Gabriel Peixoto França[2] and Nivaldo Alonso[3]

1 Amazonas State University, Manaus, Brazil

2 Alvaro Alvim School Hospital, Campos dos Goytacazes, Brazil

3 University of Sao Paulo, São Paulo, Brazil

*Address all correspondence to: joseemersonss@hotmail.com

References

[1] Atun R, De Andrade LOM, Almeida G, Cotlear D, Dmytraczenko T, Frenz P, et al. Health-system reform and universal health coverage in Latin America. The Lancet. 2015;**385**(9974):1230-1247

[2] Paim J, Travassos C, Almeida C, Bahia L, MacInko J. The Brazilian health system: History, advances, and challenges. The Lancet [Internet]. 2011;**377**(9779):1778-1797. DOI: 10.1016/S0140-6736(11)60054-8

[3] Barreto ML, Rasella D, Machado DB, Aquino R, Lima D, Garcia LP, et al. Monitoring and evaluating Progress towards universal health coverage in Brazil. PLoS Medicine. 2014;**11**:1-3

[4] Hone T, Rasella D, Barreto M, Atun R, Majeed A, Millett C. Large reductions in amenable mortality associated with brazil's primary care expansion and strong health governance. Health Affairs. 2017;**36**(1):149-158

[5] Macinko J, Dourado I, Aquino R, de Fátima BP, Lima-Costa MF, Medina MG, et al. Major expansion of primary care in Brazil linked to decline in unnecessary hospitalization. Health Affairs. 2010;**29**(12):2149-2160

[6] Santos IS, Ugá MAD, Porto SM. The public-private mix in the Brazilian health system: Financing, delivery and utilization of health services. Ciencia e Saude Coletiva. 2008;**13**(5):1431-1440

[7] Piola S, Vianna S, Marinho A. Saúde no Brasil: algumas questões sobre o Sistema Único de Saúde (SUS). Comissão Economica Para a América Latina E O Caribe - Cepal; 2009. Available from: https://repositorio.ipea.gov.br/handle/11058/1664

[8] de PF A, Giovanella L, de MHM M, Escorel S. Desafios à coordenação dos cuidados em saúde: estratégias de integração entre níveis assistenciais em grandes centros urbanos. Cadernos de Saúde Pública. 2010;**26**(2):1-13

[9] Solla J, Chioro A. Atenção ambulatorial especializada. In: Políticas e Sistema de Saúde no Brasil. [online]. 2nd ed. rev. and enl. Rio de Janeiro: Editora FIOCRUZ. 2012:546-574

[10] Wilson ML. Plano Estadual de Saúde do Amazonas 2020-2023. Secretaria de Estado de Saúde do Amazonas. 2020

[11] Instituto Brasileiro de Geografia e Estatística - IBGE. Censo Demográfico 2022 [Internet]. 2022 [cited 2023 Jul 31]. Available from: https://ftp.ibge.gov.br/Censos/Censo_Demografico_2022/Previa_da_Populacao/POP2022_Brasil_e_UFs.pdf. [Accessed: July 31, 2023]

[12] Amazonas. Plano Estadual de Saúde Amazonas 2016-2019. Secretária Estadual de Saúde 2016. pp. 1-258. Available from: http://www.saude.am.gov.br/docs/pes/pes_2016-2019.pdf

[13] Souza JE, Ferreirai RV, Saluja S, Amundson J, Citron I, Truche P, et al. Surgical capacity assessment in the state of Amazonas using the surgical assessment tool. Cross-sectional study. Revista do Colégio Brasileiro de Cirurgiões. 2022;**49**:1-9

[14] Roa L, Citron I, Ramos JA, Correia J, Feghali B, Amundson JR, et al. Cross-sectional study of surgical quality with a novel evidence-based tool for low-resource settings. BMJ Open Quality. 2020;**9**(1):1-8

[15] Massenburg BB, Saluja S, Jenny HE, Raykar NP, Ng-Kamstra J, Guilloux AGA, et al. Assessing the Brazilian surgical

system with six surgical indicators: A descriptive and modelling study. BMJ Global Health. 2017;**2**(2):1-10

[16] Alonso N, Massenburg BB, Galli R, Sobrado L, Birolini D. Surgery in Brazilian health care: Funding and physician distribution. Revista do Colégio Brasileiro de Cirurgiões. 2017;**44**(2):202-207

[17] Gawande A. Global surgery. The Lancet. 2015;**386**(9993):523-525. DOI: 10.1016/S0140-6736(15)60764-4

[18] Meara JG, Leather AJM, Hagander L, Alkire BC, Alonso N, Ameh EA, et al. Global surgery 2030: Evidence and solutions for achieving health, welfare, and economic development. The Lancet. 2015;**386**(9993):569-624. Available from: https://linkinghub.elsevier.com/retrieve/pii/S014067361560160X

[19] Implementation Tools | lcogs. Available from: https://www.lancetglobalsurgery.org/implementation-tools. [Accessed: July 19, 2020]

[20] Anderson GA, Ilcisin L, Abesiga L, Mayanja R, Portal Benetiz N, Ngonzi J, et al. Surgical volume and postoperative mortality rate at a referral hospital in Western Uganda: Measuring the lancet commission on global surgery indicators in low-resource settings. Surgery (United States). 2017;**161**(6):1710-1719. DOI: 10.1016/j.surg.2017.01.009

[21] Hsia RY, Mbembati NA, MacFarlane S, Kruk ME. Access to emergency and surgical care in sub-Saharan Africa: The infrastructure gap. Health Policy and Planning. 2012;**27**:234-244

[22] Lebrun DG, Chackungal S, Chao TE, Knowlton LM, Linden AF, Notrica MR, et al. Prioritizing essential surgery and safe anesthesia for the Post-2015 development agenda: Operative capacities of 78 district hospitals in 7 low- and middle-income countries. Surgery (United States). 2014;**155**(3):365-373. DOI: 10.1016/j.surg.2013.10.008

[23] Albutt K, Punchak M, Kayima P, Namanya DB, Anderson GA, Shrime MG. Access to safe, timely, and affordable surgical Care in Uganda: A stratified randomized evaluation of Nationwide public sector surgical capacity and Core surgical indicators. World Journal of Surgery. 2018;**42**(8):2303-2313. DOI: 10.1007/s00268-018-4485-1

[24] Datasus - Informações de Saúde. Ministério da Saúde. 2019. Available from: http://www2.datasus.gov.br/DATASUS/index.php?area=02. [Accessed: August 13, 2023]

[25] Scheffer M, Cassenote A, Guerra A, Guilloux AGA, Brandão APD, Miotto BA, et al. Demografia Médica no Brasil 2020. Conselho Federal de Medicina. 2020;**4**:264

[26] Scheffer M et al. Demografia médica no Brasil 2015. Faculdade de Medicina da USP. Conselho Regional de Medicina do Estado de São Paulo. Conselho Federal de Medicina: Departamento de Medicina Preventiva; 2015

[27] Birolini F, Saad R. Surgery in Brazil. Archives of Surgery. 2002;**137**(3):352-358

[28] Scheffer M, Cassenote A, Guilloux AGA, Miotto BA, Mainardi GM. Demografia Médica no Brasil 2018. 2018. p. 286. Available from: www.portalmedico.org.br

[29] Scheffer MC, Guilloux AGA, Matijasevich A, Massenburg BB, Saluja S, Alonso N. The state of the surgical workforce in Brazil. Surgery (United States). 2017;**161**(2):556-561

[30] Gyedu A, Stewart B, Gaskill C, Boakye G, Appiah-Denkyira E, Donkor P, et al. Improving benchmarks for global surgery: Nationwide enumeration of operations performed in Ghana. Annals of Surgery. 2018;**268**(2):1-16

[31] Watters DA, Guest GD, Tangi V, Shrime MG, Meara JG. Global surgery system strengthening. Anesthesia and Analgesia. 2018;**126**(4):1329-1339

[32] Barros FC, Matijasevich A, Maranhão AGK, Escalante JJ, Neto DLR, Fernandes RM, et al. Cesarean sections in Brazil: Will they ever stop increasing? Revista Panamericana de Salud Publica/ Pan American Journal of Public Health. 2015;**38**(3):217-225

[33] Gibbons L, Belizán JM, Lauer JA, Betrán AP, Merialdi M, Althabe F. The Global Numbers and Costs of Additionally Needed and Unnecessary Caesarean Sections Performed per Year: Overuse as a Barrier to Universal Coverage. World Health Report (2010) Background Papers; 2010. Available from: https://www.researchgate.net/publication/265064468_The_Global_Numbers_and_Costs_of_Additionally_Needed_and_Unnecessary_Caesarean_Sections_Performed_per_Year_Overuse_as_a_Barrier_to_Universal_Coverage_HEALTH_SYSTEMS_FINANCING

[34] Wise J. Target rates for caesarean section may be too low, say researchers. British Medical Journal. 2015;**351**:1-10

[35] Cristina Lima A, Cássia Januário M, Thiago Lima P, de Moura Silva W. Datasus: O Uso Dos Sistemas de Informação Na Saúde Pública. Available from: http://hdl.handle.net/10438/4873

[36] WHO. Monitoring The Building Blocks Of Health Systems: A Handbook Of Indicators And Their Measurement Strategies B WHO Library Cataloguing-in-Publication Data. World Health Organization; 2010. Available from: https://resources.relabhs.org/resource/monitoring-the-building-blocks-of-health-systems-a-handbook-of-indicators-and-their-measurement-strategies/

[37] Coelho Neto GC, Chioro A. After all, how many nationwide health information systems are there in Brazil? Cadernos de Saúde Pública. 2021;**37**(7):1-15

[38] Massuda A, Hone T, Leles FAG, De Castro MC, Atun R. The Brazilian health system at crossroads: Progress, crisis and resilience. BMJ Global Health. 2018;**3**(4):1-8

[39] Brasil. Ministério da Saúde. Fundo Nacional de Saúde. Gestão Financeira do Sistema Único de Saúde [Internet]. Ministério da Saúde. Fundo Nacional de Saúde. Gestão Financeira do Sistema Único de Saúde: manual básico. 2003. p. 66. Available from: https://www.fns2.saude.gov.br/documentos/Publicacoes/Manual_Gestao_Fin_SUS.pdf

[40] Mohanty SK, Srivastava A. Out-of-pocket expenditure on institutional delivery in India. Health Policy and Planning. 2013;**28**:247-262

[41] Nguyen H, Ivers R, Jan S, Martiniuk A, Pham C. Catastrophic household costs due to injury in Vietnam. Injury. 2013

[42] Meremikwu MM, Ehiri JE, Nkanga DG, Udoh EE, Ikpatt OF, Alaje EO. Socioeconomic constraints to effective management of Burkitt's lymphoma in South-Eastern Nigeria. Tropical Medicine and International Health. 2005

[43] Groen RS, Sriram VM, Kamara TB, Kushner AL, Blok L. Individual and community perceptions of surgical care in Sierra Leone. Tropical Medicine and International Health. 2014;**19**(1):107-116

Chapter 6

Perspective Chapter: The Economics of Biodiversity and a Pragmatic Science for the Development of the Amazon in the 21st Century

André Luis Willerding

Abstract

The use of knowledge of nature based on science and technology should be the way to economically and socially just inclusion in a new model of development of the region, generated from local conceptions but integrated with the new demands of current and future societies of the planet. Therefore, it is necessary to improve a new economy based on innovation and the sustainable use of biodiversity to generate wealth for the Amazon region and avoid activities such as cattle ranching, soy and illegal logging. The Amazon is facing a major challenge between continuing to grow at relatively low rates with a high cost to the environment and a deep social depression, or changing this pattern of development and pursuing environmentally sustained and inclusive economic growth to ensure the supply of environmental goods and services on which the well-being of the planet depends.

Keywords: bioeconomy, social development, Amazon, sustainability, biodiversity

1. Introduction

Fostering a new economy based on innovation and sustainable use of biodiversity can generate up to five times more wealth for the Amazon region than the activities currently practiced in the biome, such as livestock, soy, and logging [1]. What should be sought for the Amazon is a proposal for a new model of economic development generated from local needs and conceptions but integrated with the new demands of the current and future societies of the planet. The idea is to stop deforestation with the application of knowledge of nature to society and create an economic system based on this knowledge and innovation for the various products or services that biodiversity offers.

Currently, the Brazilian economy is very dependent on global trade in agricultural commodities and minerals, with an intensive carbon emission, high consumption of pesticides, and a low value-added agriculture. The situation gets even worse when this system reaches the Amazon, which looks for cheap land and increases deforestation and fires, promoting the emission of greenhouse gases.

With regard to decarbonization, a characteristic of the bioeconomy, Brazil can lead this global process, and the reality and necessity impose this new paradigm. Decarbonization refers to the reduction in the use of direct and indirect derivatives of fossil fuels, as well as the reduction of methane emissions in the production of goods and services. When compared with other parts of the world, the Amazon presents itself with good conditions for the so-called green economy, since it has the largest continuous area of tropical forest in the world, an abundance of water, sunlight, and a forest-based economy with great potential to grow.

However, it is not a simple and trivial discussion to try to bring up these medium- and long-term issues. After all, what kind of economy will the Amazon have in order not to miss the Fourth Industrial Revolution underway in the twenty-first century? [2] Although to discuss this, it is still necessary to solve the mishaps of incomplete or imperfect modernization that presents itself with the lack of infrastructure and logistics, for example. Therefore, what should be discussed is how to take advantage that the Amazon region has, especially in Brazil, to lead this bioeconomy, with the integration between crops, livestock and forest with their non-timber products.

Today, the world context is shifting from mass production to specialization, with the growth of agricultural products for specific market niches and with the use of biomaterials, nanotechnology, and biotechnology. At the same time, the growing consumption for the "natural," toward organic agriculture and biotechnology in addition to the use of other natural resources with unique characteristics, also encourages this transition.

What's really driving this paradigm shift? The transition from mass production technologies (including petroleum and its derivatives) to an age of "smart" information and communication technologies that are transforming all aspects of society [3].

From an economic and financial point of view, this specialization of production, unlike mass production, achieves higher profit margins. For this, science, technology, and innovation will be crucial to achieve this goal.

Thus, as this transformation spreads, consumer expectations also change, and this creates immense potential for the region.

Currently, there is a strong demand to find a solution to export "exotic" tropical fruits, develop nutraceutical plants and foods, identify medicinal virtues in little-known plants, and obtain molecules of very high value (fine chemistry) or new processes from green chemistry. These new parameters can be obtained with scientific advances that can open new horizons for local companies and innovators [4].

Biotechnology is another area ripe for innovation. They are identifying and developing microorganisms to serve many purposes, such as biological control, promotion of plant growth, biomining or digestion of oil spilled into soil or water, or applications in the area of health in the search for new active principles, not to mention the potential of fine chemistry from this microbiome. All of these changes are happening in the midst of an environmental crisis. Learning how to manage natural resources and use them to create wealth, reduce poverty, and promote a sustainable economy is the urgent task.

"Smart" green technologies are already being developed from new renewable energy sources to the creation of renewable materials. But for this to become a reality, that is, prioritizing solutions as part of a new "smart green" economy, a broad consensus is required, with bold and imaginative public policies that facilitate this transformation. It is necessary to use natural goods in the light of technologies.

From this conception, the energy sector, agribusiness, biomaterials, and the non-timber exploitation of the forest are key sectors in the transition to a green economy in the Amazon. One of the ways to discuss this is the new way to consolidate a low-carbon economy with the possibilities of new industrial sectors—bioindustries.

However, one of the major obstacles to the transition to a green economy is the lack of a regulatory framework on the use of environmental services and tangible biodiversity assets such as carbon pricing and emissions certification. As for access to the genetic heritage of biodiversity, in Brazil, the new legal framework for biodiversity, embodied by Law No. 13,123, of May 20, 2015, and Decree No. 8772, of May 11, 2016, dictates the rules today. These measures, in general, can generate value along the respective production chains.

However, while part of the world invests in a new industry that produces more, saves energy, and thus causes less environmental impact, Brazil seems to remain stuck in this transition, as it has a technologically outdated manufacturing park and a tax burden that makes national production less competitive with the outside world and curbs the investments needed to leverage innovation.

One of the ways to incentivize investment to finance "green" industry is to tax carbon emissions. Activities that pollute less will be more competitive and will have greater financing possibilities for this transition. The bioeconomy in the Amazon is moving in this direction, but as long as the country bets on non-value-added commodities, few benefits will be generalized to society at large.

However, this scenario has changed in recent times due to the speed of technological advances and changes in business models, driven by factors such as economic and political crises, the environmental crisis, and the growth of the so-called digital economy. In other words, there is more desire to make this transition through opportunities than through threats to the financial health of companies. An example is the cosmetics industry, which seeks to exchange synthetic inputs for natural substances from its inputs [5].

The green economy can reduce both environmental risks and production costs. When tax incentives occur, the tendency is to increase the profit margin. This will cause positive impacts for companies (image, for example) in addition to repositioning companies through the production of new products and services. Therefore, the use of clean technology and green chemistry can be new sources of income. For the transition to a green economy, there is a need for public policies, a lot of innovation, and market instruments to make these products more viable and attractive. Thus, the low-carbon economy agenda is also an efficiency and productivity agenda.

Thus, directing the public policy of science, technology, and innovation for the economic, social, and environmental development of the Amazon involves bringing science to society adapted to a local reality and with high economic potential, which can serve as a portal to a pragmatic science based on the need to revert to society the use of knowledge in income generation.

2. The role of bioeconomy in the social, economic, and environmental development of the Amazon

For the real development of the Amazon, it will be necessary to apply the knowledge of nature based on science and technology as the path to an economic, environmental, and socially just inclusion of the region. The great challenge will be to produce sustainable technologies, train people to use the technologies, seek ways to conserve the forest, and, at the same time, leverage a bioeconomy as a tool for the rational use of its biodiversity. To transform scientific and traditional knowledge into technological innovation, through the formation of networks, in partnership with local actors, for social inclusion, economic development, and conservation of the forest and aquatic environment of the Amazon.

The great demands in the bioeconomy of the region go through the search for solutions to the regional structural problems important for the development of the Amazon, involving a holistic look at the health of the man of the forest, the rational use of natural resources, and the training of the young people of the twenty-first century aiming at the use of the forest with the maintenance of its environmental services. Thus, the goals and proposals elaborated should meet the current discussions between the countries that hold the forests and the donors of resources in how to treat the relations between the preservation of biomes and the use of land in a sustainable way for economic and social development in addition to the need to maintain the global temperature target. And as a premise, to strengthen relations with all Amazonian actors in the search for the common good for the region..

3. Why the Amazon?

Knowledge of Amazonian nature based on science and technology is the path to economic, environmental and socially just inclusion. Producing sustainable technologies, training people to use the technologies, and seeking ways to conserve the forest and rivers should serve as indications to leverage a bioeconomy as a tool for the rational use of its biodiversity. With this, the Amazon intends to become a global reference for technological innovation through the use of the information contained in the forest. As a great challenge, to transform scientific and traditional knowledge into technological innovation, through the formation of networks, in partnership with local actors, for social inclusion, economic development, and conservation of the forest and its aquatic environment..

The continental Amazon is the last large continuous surface of tropical forests on the planet; its area of 7.9 million km^2 represents 5% of the planet's land surface. The Amazon basin covers about 7.0 million km^2, of which 5.5 million km^2 is covered by forest and most of it is in Brazil, with 60% of the rainforest, followed by Peru with 13% and Colombia with 10% in addition to Bolivia, Ecuador, French Guiana, Guyana, Suriname, and Venezuela. This Amazon rainforest represents one-third of the world's tropical forests, in addition to containing more than half of the planet's biodiversity [6].

Humanity has been relating to this region for a long time due to its riches and potentialities that contribute to the most diverse purposes. But greed has ravaged her as a disastrous way of seeing and exploiting her. Under this aspect, the need to understand how to work with the forest and its biodiversity, with water and with the Amazonian man, comes to the fore, which implies several social, environmental, and economic aspects.

The bioeconomy will have an important role in the twenty-first century, and there is in the region a great wealth that will provide sufficient material for research, studies, drugs, food, and cosmetics in a molecular universe practically still unknown, in addition to the mineral potentialities. That is, this diversity of natural resources is a real treasure, which emerges among the global environmental issues along with climate change, with the devastation of the forest for monocultures and livestock in addition to the pollution of aquatic environments, by illegal mining and lack of basic sanitation.

The maintenance of the "standing forest" in the Amazon is fundamental for nature and responsible for the balance of ecosystems and maintenance of the hydrological cycle on a global scale, in addition to the conservation of biodiversity being important for the economic potential it represents. The most consistent argument in its favor is aimed at halting the process of deforestation. This becomes extremely difficult if the Amazonian populations are not offered alternatives of quality of life superior to those that are available today and that degrade the environment. Therefore, the conservation and rational use of natural resources constitute the starting point for environmentally correct, socially just, and economically viable development.

For this, it is necessary to know the complexity of the different Amazonian ecosystems, as well as basic studies on the characterization, economic valuation, and rational use of their biodiversity.

On this aspect of tension, the Amazon is inserted as an important pole of natural resources, making the technical–scientific development fundamental as a tool for the social and economic development of the region with a way to generate income and, above all, so that there is the maintenance of forest conservation, allowing the improvement in the quality of life of the populations that maintain the knowledge of nature.

4. Man in the Amazon

Although there are few studies on the origins of man in the Amazon, archaeology shows that it has been colonized by populations of hunter-gatherers and unskilled fishermen since the late Pleistocene, perhaps some 12,000 years ago. Archeological evidence shows that these populations were adapted to the tropical forest and had habits improved over time to exploit Amazonian resources, domesticate, and/or manage different plants in association [7].

Currently, 50 million inhabitants live in the Amazon Basin, including the presence of more than 300 indigenous peoples (9% of the total population) with some ethnicities still isolated [8]. With a population of this size living in an area with diverse social, environmental, and economic needs, the environmental impact that affects the entire forest is inevitable. The "Ecological Footprint" that the Amazon has been suffering from the twentieth century to the present day has no comparison with any other time in its history, and some areas are already emitting more CO2 into the atmosphere than they can absorb [9].

With the scale of deforestation increasing in Brazil, the risk of a point of no return is very high, promoting the loss of the environmental function of the Tropical Rainforest. According to official data [4], deforestation in the Amazon in Brazil presents 470,382 Km^2 and only in 2021, the deforestation rate reached 13,235 km^2 and reached the worst level in 15 years. The current situation in the region encourages loggers, ranchers, and criminals to deforest the forest and invade public areas and harm indigenous peoples, especially as illegal mining [4]. And scientific data indicate that about 17% of the Amazon rainforest has been destroyed. Research suggests that when it reaches about 20 to 25% (therefore, 3–8% more), an inflection point (point of no return) may occur to transform it into a non-forest ecosystem such as savanna [9, 10].

Colombia has lost 1.3 million hectares of forests, caused by deforestation as a result of the expansion of agricultural frontiers—illegal and illegal—colonization, illegal mining, logging, and forest fires [11]. The Peruvian Amazon lost more than 1.2 million hectares, and 2020 was the worst year in history with the devastation of 200,000 hectares of forest with deforestation of the forest and irregular mining and contamination of water with chemicals and heavy metals [12].

To change this picture of high vulnerability, the development of the region has to be directed in investments in three main types of capital: human capital (health and education), infrastructure (digital power and transportation), and companies (bio-business). The poor countries that make up the Amazon have low levels of these types of capital. On the other hand, this difficulty can be an opportunity and a potential to grow quickly with investments in each of these strategic points. Today, this growth can and should be green and digital in a low-carbon economy [13].

5. The economic and environmental value of the forest

As an environmental service, the Tropical Rainforest in the Amazon acts as a giant "air conditioner," reducing the Earth's surface temperatures and generating rain that exerts a strong influence on the atmosphere and circulation patterns on a global scale. When returned to the atmosphere, a part of this moisture generates aerial currents known as "flying rivers" that transport about 20 billion T of water daily and the Amazon River 17 billion T of water to the ocean [10].

This profusion of natural water pumps from the forest ensures daily rainfall in the biome, in addition to regulating the entire rainfall regime in Brazil and other countries in the region. The deregulation of this system by deforestation causes temperatures to be higher and dry seasons to be more prolonged [9]. The increased occurrence and period of these dry seasons is what is leading to the "point of no return" of the forest in the region.

How to prevent the Amazon rainforest from losing its ability to feed back rainfall? In a scenario of weak governance, the southern region of the Brazilian Amazon could lose 56% of its forests by 2050 [9, 10]. With the capacity to absorb carbon decreasing, the service that the Amazon rainforest provides to the planet is decreasing, and this increases the climate crisis on a global scale [14, 15].

For the Amazon, the current challenges in economic, environmental, and social issues lead to a dilemma: (1) continue growing at relatively low rates with a high cost to the environment and a deep social depression or (2) change this pattern of development and seek sustained and inclusive economic growth to ensure the provision of environmental goods and services on which the development of a country and the well-being of its society depend [16]. The bioeconomy can contribute substantially to sustainable development by increasing the quality of the raw material produced by the forest, food security, and environmental health [17].

Within this context, it is important to emphasize and discuss the Amazon bioeconomy on the agenda of a "Pan-Amazonian Policy of Payment for Environmental Services" that will allow, in a transnational way, the actors of the value chains to adopt ecological actions in the management of the forest and its environmental resources in exchange for resources for their economic activities. This concern with ecosystem services has always been seen as an obstacle to investments, but it is necessary to reverse this scenario by adopting ecological actions in exchange for financial resources.

Starting an effective transition to a green economy requires the participation and investment of the public and private sectors in different areas, including agriculture, energy, forestry, tourism, transportation, manufacturing, and city infrastructure. Some of the investments to be encouraged are renewable energy technologies, energy efficiency in homes and bioindustry facilities, water reuse, wind energy, and efficient means of transport that converge to the valorization and maintenance of ecosystem services. This lack of attention to these markets possibly reflects the reluctance of governments to be seen as players in this shift. One of the challenges is to adapt the complexity of the value chains in this transition with the fossil economy [18].

In this sense, the viable economic activity for the Amazon goes through a decentralized bioeconomy with a network of industries and teaching and research institutes acting in technological innovation and adding value in the processing of non-timber products [15]. However, the reality shows the lack of knowledge of the mechanisms of a forest-based economy that takes into account an operational technical need for the use and conservation of the forest and its systemic environmental services of the forest. This is the challenge for the twenty-first century Amazon.

6. Strategies

The production of knowledge is based on the fact that one is studying systems consisting of a large number of agents, which integrate to produce adaptive survival strategies for the components of the system and for the system as a whole. The current demand is to work on a management model with shared and collaborative public–private partnership, in order to promote interaction between the various actors, institutes of science and technology, research and development centers, and local, national, and international academies, adding efforts for the development of strategic areas of the Amazon bioeconomy. Thus, the actions must have well-defined focuses and consider that there is a system consisting of elements that interact with each other and that is self-arranging for its sustainability and achievement of its goals. This requires the observance of the phenomenon to be studied holistically and multidimensionally [4].

An important issue is to encourage a management of research, development, and innovation in strategic areas considering the critical mass available, the local research infrastructure, the graduate programs, and the direction of the solutions of the existing "gaps" for the production of a chain of knowledge that develops its processes and products from beginning to end, aiming to meet the specific demands of the market and, above all, that fixes man and strengthens a forest-based economy.

Thus, strategic partnerships between the various actors of the region must occur to generate and support enterprises, production of goods and services, innovative solutions, contribution to the quality of critical mass, technological development, and industrial and intellectual production, bringing together public and private institutions. These are necessary bases of action to solve various problems of the populations and make the Amazonian bioproducts more competitive and with market potential.

Another important action is to attract investments for the training of human resources and contribution to training programs, development, and training and fixation of specialized professionals (technicians, specialists, masters, doctors, managers, and entrepreneurs) who carry out activities in support institutions, government, science and technology institutions, companies, and investment institutions, among others, establishing mechanisms to supply critical mass that respond to specific demands.

The training of the young people who inhabit the various cities of the Pan-Amazon should be the main factor for the conversion of the logic of forest use. To this end, interactions with science and technology institutes are extremely important for the academy to respond with the training of People and Innovation, Technology and Market so that there is social and economic growth in the region in a rational and universal way.

The purpose of this discussion is the construction of a strategic agenda that will define the actions and priorities in the bioeconomy in the Amazon. The starting point for the development of the work is the elaboration of the demands anchored in the knowledge of those who work directly with the value chains in the Amazon. For this, it is necessary to strengthen and expand the information bases to suggest lines of actions/projects/research related to technological and market demands to strengthen the main value chains. The dialog to be conducted between different sectors, including public policy makers, academia, civil society, and the private sector will serve as a panorama for various actions that can be carried out by the public power and/or the private initiative and with the inclusive participation of local society. Such answers will come from a multidisciplinary view on the various themes and issues pertinent to the bioeconomy of the region. This electronic form will serve as the first tool for guiding the priorities of the various issues and themes to be addressed.

7. Human resources training

Involve high school, technical, and technological institutions, in network, in the formation of young people for the Pan-Amazon of the twenty-first century to contribute to the design and implementation of professional training models appropriate to Amazonian issues. For this, it is important to strengthen and expand training and extension programs for the sustainable development and conservation of the Pan-Amazon and to facilitate the mobility and exchange of researchers and students participating in activities related to the sustainable development of the Amazon. Another challenge is to promote mechanisms to stimulate the establishment of qualified personnel for scientific and technological production in the Pan-Amazon, with emphasis on strategic sectors in line with regional, national, and international cooperation, which enables the creation of thematic networks, minimizes efforts, and stimulates integrated scientific initiatives for the study of the environments and biodiversity of the Amazon.

The creation of new teaching modalities and innovative approaches on biodiversity with the strengthening of entrepreneurship, as well as the creation of transdisciplinary poles and educational centers of excellence, demands large-scale public–private financial incentives. The development of the bioeconomy depends on new technologies based on local knowledge. For this reason, the principles of sociodiversity in the Amazon are fundamental to begin the transition to a sustainable economy. In parallel, education must remain connected with the global ST&I ecosystem and ensure cognition processes for Amazonian students. These educational models can provide local talents with the competencies to meet the demands of each micro-region and international markets [13].

8. Training of young people from the interior of the Amazon region for the needs of the Bioeconomy

Prepare high school students for jobs in the region and meet the training needs of young people, allowing them to undertake for the development of the bioeconomy and support state networks and federal institutes present in the Legal Amazon to implement technical and professional training focused on the main value chains of the bioeconomy..

9. Strengthening the value chains of the bioeconomy in the Amazon

The components of the production chains will be analyzed locally, and the inhibitory and high-performance driving factors will be identified to propose solutions that respond to the identified problems. The method used will be the case study, analyzed with a multidisciplinary view, considering the three units of analysis proposed in the agreement: açaí, Brazil nuts, and guarana, but other important units such as cocoa and pirarucu may be inserted. This primary analysis of the situation aims to identify critical factors to increase the competitiveness of productive chains with consumption potential in the markets and for the generation of employment and income.

10. Generation of sustainable biobusinesses in the Amazon

Sustainable practices bring competitive advantage to Brazil and the national productive sector, which must be increasingly committed to an inclusive economic model and that is also established with the inclusion of an Amazon bioeconomy. The business from the forest assets in the Amazon encourages social and environmental transformations, with the generation of work and income for Amazonian communities, maintaining the "Forest Standing" with reduction of deforestation, biodiversity conservation, and the maintenance of an economically active population in the interior, fostering the forest-based economy.

11. Challenges for the implementation of sustainable biobusinesses in the Amazon

Challenges to implementing sustainable biobusinesses in the Amazon include the lack of infrastructure and logistics, the lack of fiscal and financial incentives, the lack of technical and managerial training of producers, and the lack of market access. In addition, deforestation and burning are a major problem in the region.

12. Conclusion

When it comes to the Amazon region, it is observed that there is no simple option for the development of the region. The challenge of combining the necessary economic development with the preservation of the forest implies building paths capable of generating income and quality of life for their populations. This quest for development must lead to an interaction between social forces capable of using the riches derived from biodiversity and other regional natural resources without destroying them. And this cannot be achieved by replicating the current patterns of economic development.

The main challenge in the development of the region is the formation of a critical mass with the capacity to analyze and provide solutions to Amazonian problems. An essential condition to be used will be the critical capacity of educational and research institutions in the Amazon to develop local collaborative activities to understand the dynamic processes of the Amazon biome and stimulate new businesses involving biodiversity and environmental services.

The proposal described here serves the Amazon region, and its development requires exactly something that enhances the transformations and induces really enabling options for a more promising future, as well as promotes greater articulation of regional structures to national and international circuits, in order to intensify trade and financial flows and cultural, scientific, and technological exchange.

Author details

André Luis Willerding
Ananas Biofactory (Biofábrica Ananas), Manaus, Amazonas, Brazil

*Address all correspondence to: alwillerding@gmail.com

References

[1] Abrantes JS. Bio (sócio) diversidade e Empreendedorismo Ambiental na Amazônia. Rio de Janeiro: Garamond; 2012

[2] Consultoria B. Strategic Roadmap for the Brazilian Bioeconomy. São Paulo: Natura, Givaudan e Kimberly-Clark; 2017. p. 26

[3] Viana D. Riqueza que vem da floresta. Revista Fapesp. 2021;**306**:77

[4] Willerding AL et al. Estratégias para o desenvolvimento da bioeconomia no estado do Amazonas. Estudos Avançados. 2020;**34**(98):145-166. DOI: 10.1590/s0103-4014.2020.3498.010

[5] Sousa KA. A dinâmica da inovação em bionegócios no estado do Amazonas. Belém: XXIV Seminário Nacional de Parques Tecnológicos e Incubadoras de Empresa; 2014. p. 25

[6] OTCA -http://otca.org/pt/a-Amazônia/. A Amazônia. Acesso em Novembro 2021

[7] Cassino MF et al. Arqueobotânica dos Povos Indígenas Brasileiros e Suas Plantas Alimentícias. In: Jacob MCM, Albuquerque UP, editors. Local Food Plants of Brazil. Etnobiologia. Cham: Springer; 2021. DOI: 10.1007/978-3-030-69139-4_8

[8] ISA. Povos indígenas no Brasil - Línguas. Instituto Socioambiental; 2020. Available from: https://pib.socioambiental.org/pt/L%C3%ADnguas. [Accessed: July 2020]

[9] IPAM. https://ipam.org.br/pt/ Instituto de Pesquisa Ambiental da Amazônia – acessado em dezembro de 2021.

[10] Silva Junior CHL et al. Amazônian forest degradation must be incorporated into the COP26 agenda. Nature Geoscience. 2021;**14**:634-635. DOI: 10.1038/s41561-021-00823-z

[11] Minambiente. Estratégias abrangentes para o controle do desmatamento e manejo florestal na Colômbia. 2017. Ministério do Meio Ambiente da Colômbia MINAMBIENTE. 37. pp. 3. García Romero, Helena. In: Desmatamento na Colômbia: desafios e Perspectivas. Fedesarrollo; 2014. p. 28

[12] MAAP. https://climainfo.org.br/tag/relatorio-projeto-de-monitoramento-da-Amazônia-andina-maap/.Acessado em dezembro de 2021

[13] Sachs JD. Time to overhaul the global financial system. Project Syndicate. 2021.Available from: https://www.project-syndicate.org/commentary/global-financial-system-death-trap-for-developing-countries-by-jeffrey-d-sachs-2021 [Accessed: December 2021]

[14] Nobre C. Amazônia 4.0: The Amazon Third Way Initiative Brief. April 2021. 5p. Available from: http://amazoniaquatropontozero.org.br [Accessed: December 2021]

[15] FAPESP. Os Limites da Amazônia. Novembro de. Ano 20, 2019;**285**:32-39. Available from: www.revistapesquisa.fapesp.br [Accessed: November 2019]

[16] Lokko Y et al. Biotechnology and the bioeconomy – Towards inclusive and sustainable industrial development. New Biotechnology. 2018;**40**:5-10

[17] Philp J. The bioeconomy, the challenge of the century for policy

makers. New Biotechnology. 2018;**40**:11-19

[18] Homma AKO. Extrativismo vegetal ou plantio: qual a opção para a Amazônia? Estudos Avançados. 2012;**26**(74):167-186